D1118588

Praise for *OPEN*

"An upstart company in Texas, with cows grazing outside its windows, bets its future on open standards, only to find itself battling the world's most powerful technology company for control of the personal computer industry. Rod Canion reveals the back-room battles, secret alliances, and bet-the-company decisions he made as CEO of Compaq, which he guided from start-up to Fortune 500 in less than four years. Canion's process for making executive decisions will be of interest to managers in any competitive industry."

—PETER H. LEWIS,
former senior writer and technology columnist, The New York Times

———

"Compaq's early business decisions changed the course of personal computing. This is a detailed, inside look at those high-risk, high-reward calls by the executive who made them and holds important lessons for competitive strategy."

—RICHARD SHAFFER,
former technology columnist for The Wall Street Journal,
Forbes, *and* Fortune

"Few technical arcana put laymen to sleep faster than a discussion of industry standards, even though winning and losing in technical business often depend on them. Canion's description of the human side of cobbling together what's needed to create one of these standards is correct—and a good read, too."

—ANDY GROVE,
former CEO of Intel

———

"David was a lot smaller than Goliath, but his relative size paled in comparison to that of start-up Compaq Computer when it set out in 1982 to overtake personal computer giant IBM. Yet remarkably, just a decade later, Compaq had successfully toppled IBM as the world's largest PC company. In this fast-paced recounting of how the inconceivable became the actual, Compaq cofounder Rod Canion tells how daunting hurdles were overcome and opportunities seized. The press, the analysts, the Wall Streeters had said at the beginning it couldn't be done. Rod and his team did it. *Open* will take you along on this exhilarating ride through technology, innovation, and unprecedented industrial growth."

—BEN ROSEN,
former Chairman of Compaq

———

"*Open* is a fascinating insider account of how the once IBM-dominated proprietary computer industry was transformed into one of open standards in the late 1980s and early 1990s as upstart startup Compaq, with a little help from Intel and Microsoft, decisively defeated IBM on Big Blue's own turf. IBM's crushing defeat in the PC market set a crucial precedent for the replacement of traditional proprietary architectures with open standards for systems, networking, and software platforms across the information industry.

It is always tempting, looking back at history, to assume the inevitability of whatever actually happened. Canion's inside account the founding of Compaq, its success, and its crucial role in the ultimate victory over IBM in defining an industry standard PC architecture makes it clear that the outcome would have been very different, but for crucial (and often risky) choices by Compaq.

Had Canion and his colleagues at Compaq been less bold (or executed less well), the shape of the information industry today might be radically different; still be dominated by traditional proprietary players like IBM and AT&T. In crushing IBM's attempt to lock back up the PC market with proprietary architectures, Compaq tumbled the first of a series of dominoes that opened the way to today's still vibrant and growing markets built on quality and innovation around open standards. *Open* is also, however, a valuable account of one of the most successful startup companies of the early 1980s with important lessons for startup companies today. This book is a must-read for anyone seriously interested in innovation, investment in startups, or the information industry."

—WILLIAM F. ZACHMANN,
Computer and Communications Industry Analyst and former senior VP of Market Research at International Data Corp.

OPEN

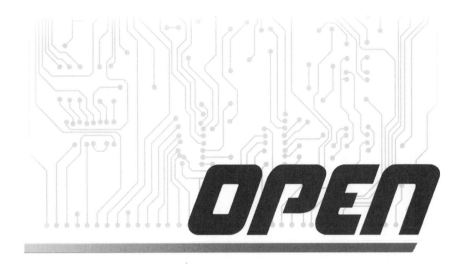

OPEN

How Compaq Ended IBM's PC Domination
and Helped Invent Modern Computing

ROD CANION

BENBELLA

BENBELLA BOOKS, INC.

Dallas, Texas

BenBella Books, Inc.
10300 N. Central Expressway, Suite 530
Dallas, TX 75231
www.benbellabooks.com
Send feedback to feedback@benbellabooks.com

Printed in the United States of America
10 9 8 7 6 5 4 3 2 1

Library of Congress Cataloging-in-Publication Data
Canion, Rod.
    Open : how Compaq ended IBM's PC domination and helped invent modern comput-ing / Rod Canion.
        pages cm
    Includes bibliographical references and index.
    ISBN 978-1-937856-99-1 (trade cloth : alk. paper)—ISBN 978-1-936661-92-3 (elec-tronic) 1. Compaq Computer Corporation. 2. Computer industry—United States—History. 3. Microcomputers—United States—History. 4. IBM microcomputers—History. 5. COMPAQ Portable Computer—History. I. Title.
    HD9696.2.U64C653 2013
    338.7′62139160973—dc23   2013022597

Editing by Kenneth Kales and Russell Setzekorn
Copy Editing by Dorianne R. Perrucci
Proofreading by Stacia Seaman, Vy Tran, Greg Teague, and Chris Gage
Cover design by Faceout Studio
Text design and composition by John Reinhardt Book Design
Printed by Bang Printing

Distributed by Perseus Distribution
To place orders through Perseus Distribution:
Tel: (800) 343-4499
Fax: (800) 351-5073
E-mail: orderentry@perseusbooks.com
www.perseusdistribution.com

Significant discounts for bulk sales are available. Please contact Glenn Yeffeth at glenn@benbellabooks.com or 214-750-3628.

This book is dedicated to the thousands of men and women who helped make the Compaq dream into an unprecedented success. After you strip away all the trappings, it's the people of a company who either make it or break it. The Compaq team made it like none before.

Contents

Prologue
Why the Open Industry Standard Matters

THE PHRASE "open industry standard" isn't used much today, but from late 1983 until the early '90s, it was a central theme of the personal computer (PC) industry. Then it became a permanent part of the culture, and, gradually, almost everyone forgot about it. But whether we realize it or not, many of the conveniences we take for granted today exist because of the impact of the PC open industry standard on computers and technology. If it hadn't existed or if it had ended prematurely, we would be doing things a lot differently.

Take, for example, the automobile industry. In the beginning, cars worked differently. New buyers had to be taught how to drive a particular car—and if they changed brands, they had to learn a different way to drive. Then most automakers began to standardize the way they built cars. Steering was done by a wheel in front of the left front seat; the "gas," or accelerator, was under the right foot; the brake pedal to the left; and the clutch pedal to the left of the brake.

With standardization, once a person learned how to drive any brand of car, he could change brands and not have to learn again. Other features became standardized: all cars began to use gasoline so pumps along the road featured the same type of fuel; spaced their tires the same width, so roads could be built; and used similar materials, so that suppliers could gear up for higher volumes and reduce costs.

Since a buyer could switch brands easily, competition became more intense and prices came down. The result was that growth in the auto industry exploded and technology advanced rapidly.

Before the PC industry standard, the computer industry was similar to the early automobile industry, but once standardization began, the effects went much deeper. Industry-standard PCs worked the same, their keyboards and screens looked the same, and their diskettes loaded and unloaded the same. PCs even used the same "engine," which in a PC was the microprocessor and memory. There was also something new called "software," which was necessary for PCs to be able to do useful work. All industry-standard PCs used the same software. While cars had add-on options, PCs had add-in boards and peripherals, such as modems and printers.

This meant that the companies that produced the parts, software, add-in boards, and peripherals could focus all their resources on a single version instead of five or ten different ones as they had done in earlier computer markets. In addition, when they came up with a new capability or innovation, they had to implement it only once for it to work on all industry-standard PCs.

By the late 1980s, all PC technologies began to advance more rapidly because of this increased focus and the increased competition. The focus led to more efficiency, while the competition provided the incentive and motivation. The third ingredient was capital, which was more available because of the explosive growth of the market.

Over the next two decades, processor performance and memory capacity grew more than a thousandfold. Screens went from bulky, heavy, and power-hungry monochrome cathode ray tubes to thin, light, and low-power color liquid crystal displays (LCDs). Data storage expanded from 40-million-byte disk drives to 4 terabytes—a 100 thousandfold increase.

The magnitude of these advances is hard to comprehend today. But while technology would have advanced without the industry standard, it wouldn't have advanced at the same staggering rate.

The iPhones and iPads so ubiquitous today could not have been created in 1980 or 1990—they weren't even possible in the first few years of the twenty-first century. Different companies had similar ideas and made repeated attempts, but the technologies to make them easy to use weren't invented yet and the chips weren't powerful enough. The tiny computers we hold in our hands today are the result of a very long series of advances and innovations that began with Intel's invention of the microprocessor during the 1970s and finally reached critical mass in the mid-2000s. Without the increased rate of technological advance that came as a result of the open industry standard, we would still be waiting for the technology necessary to create the amazing devices we now take for granted.

The Beginning

The twenty-first-century Information Age came to life and took its first steps at a particular moment in time, over a unique eight-year period that still shapes the direction of technology today, especially personal technology rushing to fill every space of our lives. The ever-more mobile, ever-more powerful, ever-more personal information tools we hold in our hands trace their origins directly to the battles won and lost in the 1980s for the future of personal computing and technology.

If you walk into a computer store today and select any PC software package, you expect that program to run correctly on any brand of PC—whether it's a PC from Hewlett-Packard, Compaq, Dell, Sony,

Lenovo, Asus, or any other company. But if you had walked into a computer store in 1981, every brand of computer required a different, specially adapted software package, much like what exists today with different video game consoles.

What happened between 1981 and 1989 that fundamentally changed the PC industry? Most would say IBM entering the PC market caused the change, and that certainly was an important part of the story. IBM was the largest computer company by far at that time, and its brand was as strong as Apple's is today. Although IBM's PC established the open architecture around which the industry standard developed, the truth is, IBM never intended for other PCs to be able to run the same software. The computer giant didn't want standardization; IBM's executives liked the status quo. It had become the world's largest and most powerful computer company playing by the old rules—and it was willing to do almost anything to keep it that way.

IBM's executives liked the status quo. It had become the world's largest and most powerful computer company playing by the old rules.

What about Intel and Microsoft? These companies stood to gain the most from an industry standard that would require Intel's microprocessors and Microsoft's operating system. But neither of them was in a position to define and develop a standard involving every part of a personal computer. The creation of a true industry standard that would grow and evolve required a PC manufacturer other than IBM that wanted a standard to exist. However, established computer companies, such as Hewlett-Packard (HP), Digital Equipment Corporation (DEC), and Texas Instruments (TI), initially introduced proprietary PCs and were also content with the old rules. When they finally did figure out where the industry was headed, they were too late getting into the game—and someone else had already taken the lead.

Compaq Steps Up

That leader turned out to be Compaq. We had the need—and will—to achieve total compatibility. From the beginning, the company's founders were absolutely certain that we had to be able to run IBM PC software straight out of the shrink-wrapped box; otherwise, we would have no software at all for our portable PC. And we soon figured out that we would need the whole industry with us if we were ever going to get out from under IBM's shadow.

Making our portable run all IBM PC software meant we had to find and fix every single incompatibility—one at a time. It was very tedious and time consuming, which perhaps explains why no other company ever achieved the same degree of compatibility with IBM as Compaq. We didn't realize it at the time, but we were developing an incredibly important and valuable technological capability. We learned how to design PCs to include key innovations while maintaining complete backward compatibility; that is, the ability to run all existing software and use all existing add-in boards and peripherals.

Compaq quickly gained the reputation for being the most IBM-compatible of all the PC brands and eventually for even being more backward compatible than IBM. Then in late 1983, in a transaction known only to a few people, Compaq licensed our own version of Microsoft's Disk Operating System (MS-DOS) *back* to Microsoft for it to sell to our competitors. In doing so, Compaq set the industry on a course toward more consistent compatibility with an industry standard that gradually became less dependent on IBM.

As the movement toward standardized software and add-in boards gained momentum and acceptance during the mid-'80s, IBM decided to abandon the architecture that had started it all and changed to one that was much more tightly controlled and protected. It didn't want competitors to sell personal computers that could run IBM PC software without buying a license that carried significant royalties. At the time, that seemed like a smart move, one that industry observers had anticipated for a long time. Given IBM's dominant leadership position in the PC market, its strategy should have succeeded. But what IBM didn't count on was not all of its competitors were willing to surrender without a fight.

Anyone looking at Compaq during our first few years could not have guessed that we would emerge to lead the PC industry in its revolt against IBM's move toward proprietary control. We were a tiny start-up that built only portable computers—hardly a threat to giant IBM. Even when we did move into desktops, our market share was almost immeasurable at first.

But in a sequence of decisions and events that essentially amounted to a "perfect storm," Compaq quickly moved up to the third position behind IBM and Apple and developed a reputation as a technology leader and innovator, just in time to resist IBM's bold move. IBM surely believed that no company—not even Compaq—could block its success. What IBM never

How was Compaq able to move so quickly, from a start-up wannabe with a four-page business plan to an industry leader capable of defeating IBM?

expected—and what was arguably Compaq's most important decision ever—was that we would recruit and lead all the other PC companies to join us in the battle. If it was IBM versus Compaq, IBM would almost certainly win, but if it was IBM versus the entire PC industry, *that* would be a real battle.

In the end, IBM didn't succeed in taking proprietary control of the PC industry. The perfect storm that had prepared Compaq to lead the clone army continued to blow in our favor as the press gave very positive coverage to the "Gang of Nine"[1] announcement about

---

1 AST Research, Compaq, Epson America, HP, NEC, Olivetti, Tandy, Wyse, and Zenith.

our industry coalition. The successful introduction of the Extended Industry Standard Architecture (EISA) bus was followed a year later by products that proved EISA's performance superiority over IBM's Micro Channel. The industry coalition completely debunked IBM's claim that it had no choice but to sacrifice compatibility in order to achieve increased system performance.

As a result, there was a sense that IBM had tried to scam the industry. Its reputation was damaged not only in PCs, but in the rest of the company's business as well. A few years later, IBM announced that it was discontinuing production of the PS/2, and some years after that, IBM left the PC business completely.

How was Compaq able to move so quickly from a start-up wannabe with a four-page business plan to an industry leader capable of defeating IBM? The complete story includes: extraordinary people, a unique company culture, bold and often unexpected strategy, a powerful decision-making process, excellent execution, and quite a bit of good luck. Intermingled throughout our story were a handful of extraordinary decisions, each one a turning point that resulted in Compaq taking advantage of both our competitors' mistakes and the opportunities then emerging during the formation of this new personal computer industry.

To really understand why the open industry standard developed, it is helpful to go back before IBM introduced its PC and follow the sequence of events that developed. It actually happened in stages over a period of years.

Every Computer Required Different Software

Before IBM entered the PC business in 1981, every PC required a different version of every software program. A Radio Shack store would carry a proprietary version of VisiCalc, the first spreadsheet program, which ran only on the TRS-80. A ComputerLand store would carry

IBM quickly became the market leader with the introduction of its Personal Computer.

a version of VisiCalc that would run only on the Apple II, and, in addition, a different version for the Commodore 64. When VisiCalc became the most popular PC program, any PC without its own version of VisiCalc was unlikely to find shelf space in a computer store and was, therefore, likely to fail.

VisiCorp, the company that developed and sold VisiCalc, had to adapt its program to each new PC. Limited software development resources meant that not all PC brands would have VisiCalc at the same time; one or two brands would have it first, and then one or two more would follow, and so on over a period of months. That situation existed for every software application company and every PC company. It was a vicious cycle: The PC brands that got the most customers tended to get the popular programs first, and the PC brands that got the popular programs first tended to have the most customers. A fierce battle developed between PC companies to get software companies to adapt popular programs to their PCs first, or at least put them early in the pipeline.

This became a very big problem for start-ups. A major computer brand, such as HP or DEC, could get a good place in line for one of its new PCs, but a new and unknown computer company like Compaq was completely left out in the cold. The situation seriously threatened to cut off the flow of new, innovative PCs from start-up companies. Then along came IBM, the largest computer company in the world. At first IBM's entrance to the PC market made the problem worse, because almost all software and peripheral companies immediately moved IBM to the top of every development priority list. But before long, it became apparent that IBM had enabled an unexpected—and unintended—solution to the problem.

When IBM first introduced its PC, the strength of the IBM brand attracted almost every software and peripheral company. Its PC sales took off to a great degree because all the popular software and peripherals quickly became available for it. It soon became the market leader, and the software and peripheral companies continued to first introduce their new products to run on the IBM PC. Although there were still a dozen other PC brands to which they would adapt their products over the following months, start-up PC companies would never get them.

Then Compaq and a few other companies discovered the solution IBM unintentionally helped create—if software companies wouldn't adapt their programs to a new company's PCs, then new companies would adapt their PCs to run the popular software that already existed. In IBM's rush to get its PC to market, it had not taken the time to incorporate any major barriers to cloning, which is when a company makes its PC run the software written for another company's PC. Compaq and other companies set out to create PCs that would run IBM PC software; from that, the IBM PC-compatible market was born. We had effectively eliminated, or at least lowered, the barriers to entry for new and existing computer companies. At first, some established computer companies were skeptical, but as Compaq's sales grew at a staggering rate in 1983, it wasn't long before every company but Apple was jumping on the PC-compatible bandwagon.

Just five years after IBM introduced its PC, IBM and a few dozen PC compatibles accounted for more than 75 percent of the U.S. market. Apple's market share fell below 15 percent and the PC industry had been completely reconstructed around the open industry standard. But in spite of its early success, there were a few more challenges that had to be overcome before the future of the open industry standard was assured.

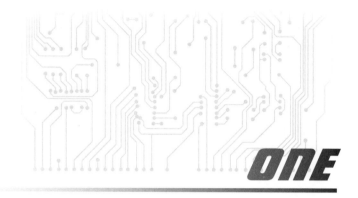

An Unlikely Beginning

It's a typically hot and sticky Tuesday evening in a Houston suburb as Jim Harris and I knock on the door of the home of Bill Murto. The three of us are longtime coworkers at Texas Instruments, and Bill welcomes us warmly, ushering us into his dining room and to seats around the table. We've decided to finally meet and have a serious discussion about starting our own company. We've been joking about the possibility for a long time, but frustration with TI's management has begun to make us take the idea much more seriously.

The three of us chat for a few minutes until Bill quickly gets to the point. "I think we should each write down our goals in starting a company. We don't want to find out later that we're trying to achieve different things."

We spend the next several minutes quietly writing our goals. Then I say, "Jim, why don't you go first."

I jot down notes as Jim says, "My goal is the personal satisfaction of succeeding in my own business. Also, I want to work in an environment where reward is proportional to effort."

As I am writing this down, Bill jumps in and says, "Rod, you're next."

"My goal is to be able to control the environment that I work in, so that the team is free to do what we believe makes sense. Plus, I'd like to achieve financial security for my family."

Jim and I look at Bill. He says, "My goal is to work in an environment of trust among my peers. I also want the personal satisfaction of creating and operating a good environment for people to work in, and I agree that monetary reward should be proportional to contribution."

After discussing our goals for a while, we shift to discussing what our business should be. We have talked in the past about designing and producing add-in boards for the new PC from IBM that is taking the personal computer market by storm. This seems to us like a relatively easy product to design and doesn't create a conflict with our current employer—something we want to avoid at any cost.

After meeting for about an hour, Jim and I head home with an air of excitement. We've taken the first step toward starting our own company, but there's no time for celebrating since we have to be at work early the next day.

It's worth noting that the primary issue discussed at our first serious meeting wasn't what product we were going to market, or who was going to be the boss, or even how much money we wanted to make. The core discussions centered on what kind of environment we wanted to create and work in—one that would be fair and make sense. We didn't know it at the time, of course, but that environment would promote values at the heart of a culture that enabled our unprecedented success. But that lay far in the future.

The formidable path to starting a company from scratch stretched out ahead of us.

It wasn't by chance that the three of us decided to start a company. Jim and I were both electrical engineers and had worked together on several different projects at Texas Instruments (TI) since 1972. I met Bill in 1977 when we both worked for TI in Austin. He impressed me with his marketing knowledge and insight. When I was asked to head a project to get TI into the Winchester disk business in 1980, I picked Jim and Bill to help me figure out what made sense for TI. During that project, the three of us were exposed to the venture capital culture of Silicon Valley, and it was a real eye-opener. We discovered, too, that we worked together well as a team, with Jim as engineering manager, Bill as marketing manager, and me as Product Customer Center (PCC) manager.

WE DIDN'T MEET AGAIN for three weeks. By then, our frustrations with our jobs had grown and we began to work on our business plan more intensely. None of us had ever created a real business plan, so I visited Houston's public library and checked out several books describing the different parts of a business plan we would have to present. We decided that our first product would be an add-in Winchester disk and controller board for the IBM PC. Each of us worked on a section, carefully following the outlines presented. When it came to projecting shipment units and revenue, we were completely at a loss. I eventually made a guess so we could finish the document.

I decided to contact L. J. Sevin, a venture capitalist in Dallas I had met through Portia Isaacson, a well-known industry consultant who had visited my group in Houston in October 1981. At the end of her visit, I offered to drive her to the airport so I could ask her about venture capital. She mentioned that Sevin was a partner in a new venture firm. I called Isaacson and asked her to set up the introduction. After I sent Sevin a copy of our plan, he showed it to his partner, Ben Rosen, who was based in New York. They said they would take a look at it and get back to us.

Jim and I were scheduled to attend one of the computer industry's largest trade shows, Comdex, in Las Vegas in mid-November. When we learned that Sevin and Rosen were going to be there, we set up a meeting in their hotel suite. We hadn't gotten any feedback from them about our business plan or their interest in investing before we met with them, so we hoped to get some news. All they had to say was that they were still looking at it. But getting caught up in the furor of the PC market at Comdex led us to decide to resign from TI so we could focus on starting our company. We still hoped that Sevin and Rosen would decide to fund us, but we didn't want to wait any longer to get started.

So we were on the verge of resigning when we ran into an unexpected problem. Bill's wife, Maura, was very close to giving birth, and he was concerned about losing insurance coverage. We decided to wait for the baby to be born before leaving our jobs. After Bill and Maura's daughter was born on December 2, there was a minor complication and Bill decided to wait to see how her condition turned out. But Jim and I, not wanting to wait any longer, resigned.

It was on the morning of December 4, 1981, that Jim and I each submitted a short letter of resignation. We gave the company a month's notice. Our department managers tried to talk us into staying, but after a while, seeing that we were committed to leaving, they terminated our employment on December 14.

While we were still employed at TI, Jim, Bill, and I were careful to not work on our new company at the office. Often we would go to Jim's house during our lunch break and work on various issues. On December 9, we were working during lunch at Jim's house and were just about ready to head back to the office when the telephone rang. It was Sevin.

He had bad news. They had talked to VenRock, a well-known venture firm, about our idea. While the people at VenRock liked it, they had a conflict due to another investment. Sevin and Rosen had also talked to Kleiner Perkins, another well-known firm, and they were negative on the idea itself. Based on that information, he said they had decided to pass. Sevin, however, made it clear that he and Rosen liked what they had seen from us and would be interested in looking at a different product idea in the future.

We were very disappointed but, strangely, not discouraged. We discussed the possibility of changing our direction and staying with TI, but neither of us wanted to give up on what we had just started. We decided to put the add-in Winchester disk product idea aside and look for other, more interesting possibilities. We realized we had been very constrained in thinking of ideas that did not conflict with TI. As soon as we were no longer employed there, there were a lot fewer constraints on what we could consider.

We decided to each look into different product areas and find a more interesting idea on which to base our start-up. We were careful to only let Bill look at areas that we believed were not a conflict with his employer, but Jim and I were free to look wherever we chose.

We picked "Gateway Technology" for the temporary name of our new company, and on December 14 reserved that name in the DBA (Doing Business As) records of the Harris County Courthouse. Then we opened a bank account and deposited $2,000, the sum of $1,000 investments from both Jim and me. That made it seem like we were really starting a company. We just didn't have a product idea yet.

Jim and I met almost daily to discuss our progress. We had a definite time frame to come up with an idea and get venture funding, or we would have to find jobs with existing companies. Our excitement about starting a company was tempered by the knowledge that we had set aside only six months of living expenses for our families.

Throughout December and early January, the three of us considered dozens of product ideas. Some were discarded right away; others we dug into more deeply. Then, on the morning of January 8, something unusual happened.

Sitting alone in my breakfast room drinking coffee, I was turning over the portable computer market in my mind and writing notes. I liked the fact that a portable wouldn't directly compete with IBM's desktop. I was looking for user needs that weren't being met and ways we could differentiate ourselves from our competitors, particularly focusing on displays, storage, and styling. I had ideas about all three areas, but I couldn't shake my biggest concern. No matter how good the product might be, a start-up wasn't going to be able to persuade software companies to adapt the most popular software applications to run on it. Without software, the product had absolutely no chance of success.

Suddenly, I was struck by a remarkable idea that was so simple and obvious it sent a chill down my spine. What if we could make a portable computer run the software written for the IBM PC? Could we design a rugged portable computer with a nine-inch display and professional styling and guarantee the best software would be available for it? The idea was so exciting, I couldn't believe I hadn't seen it before.

Then my engineering instincts kicked in. If it was so obvious, there must be some fatal flaw I was missing. I began listing the negatives:

WHY NOT DO THIS PRODUCT?
1. It's already being done.
2. Product is too heavy.
3. Can't build it cheap enough.
4. Can't get computer stores to carry it.

Nothing on the list seemed insurmountable. After a while, I called Jim to get his reaction. He liked the idea, but suggested we continue looking at other ideas too. We also decided not to mention the idea to Bill yet, since it might cause a conflict with his job at TI.

Then I went back to my notes and wrote:

PLAN OF ATTACK
1. Key issue, "Is it already being done?"
2. Talk to Ben, Portia.
3. If feedback OK, proceed toward business plan.

I had a strong feeling that this portable computer idea was the one.

Jim and I asked Sevin and Rosen to meet with us the next time they were in Houston. They agreed to meet on January 20. We had planned to go over several potential products and get their feedback on which idea looked the most promising, but the portable computer idea continued to gain momentum. So we decided to put together a short business plan and focus the meeting on convincing Sevin and Rosen to invest in it.

Writing the product description didn't take very long, but Jim and I felt words weren't enough to communicate the concept of professional styling. We decided we needed a sketch to make it clear how different our product would be from the leading portable computer in the market, the Osborne I.

Jim called Ted Papajohn, an industrial designer recently retired from TI, and asked him to draw the sketch for us. A meeting was set for 10:00 A.M. on Monday, January 18, at the ComputerLand store on Westheimer, west of downtown Houston. I showed Ted both the IBM PC and the Osborne I. The idea, I told him, was to make our product look and operate like the IBM PC. While we would arrange the display, keyboard, and floppy disk drives in a manner similar to the Osborne I, we definitely didn't want it to look like the Osborne I, which I referred to as "Army surplus."

The three of us didn't want to risk being overheard, so we walked across the street to the House of Pies restaurant for a cup of coffee. We asked for a table near the rear of the restaurant, but it really didn't matter since no one else was there at the time. After ordering, we realized we hadn't brought any paper or pencils with us. The waitress offered her pencil, but didn't have any paper. No problem—the placemats were made of paper. We turned one over and Ted began to sketch the product we had described.

When we were finished, Ted asked us when we needed the drawing. "Tomorrow," I replied, because we had a meeting in two days. Ted grumbled a bit, but said he would do the best he could.

The next day, January 19, I turned thirty-seven, but barely noticed. We were in the heat of the battle, and I was totally focused on creating a business plan that would convince Sevin and Rosen, or some other venture capitalist, to fund us to develop this product. That afternoon, Jim and I went to Ted's house to look at his drawing. We liked what he had done, so he agreed to finish detailing the drawing. We could pick it up the next morning.

Our appointment with Sevin and Rosen was set for 2:00 P.M. in a room at the Hilton Inn on North Beltway 8 that Jim had reserved for the meeting. On his way there, he picked up the finished drawing from Ted and stopped at a Kwik Kopy store to have several color copies of the sketch duplicated. Then he met me at the Hilton Inn to grab a quick lunch and prepare for the meeting.

JANUARY 20, 1982, 2:00 P.M.

A taxi pulls up in front of the Hilton Inn and drops off Sevin and Rosen. Jim and I greet them, shake hands, and lead them to a room on the first floor. I hand the two men copies of our four-page business plan and walk them through the product description, concluding with the drawing of the product. When I finish the description, I am surprised to see little reaction from Sevin or Rosen. After a tense moment, I ask them what they think.

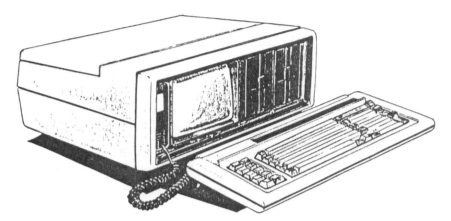

Ted Papajohn sketched a portable computer for our business plan.

Sevin responds, "The product is a no-brainer. In fact, I thought of this several months ago."

Rosen jumps in. "Actually, I thought of it before you did. Don't you remember me mentioning it to you?"

Sevin fires back, "Yeah, but I thought of it before that."

Jim and I look at each other, wondering what is going on. Finally, I ask, "If you like the idea so much, then what's the problem?"

Sevin looks at me. "Remember, you guys brought us a rather weak product idea at first, so we're wondering if you have the capability to pull off something this big."

I look over at Jim. We both smile. "Oh, that's a relief. This is the kind of product we've been doing for five years."

Jim adds, "We've been doing sturdy aluminum chassis, molded plastic enclosures, and microprocessor-based products for even longer than that."

The four of us discuss different aspects of designing such a product, and then start discussing how to make our product run IBM PC software.

I say, "I think the ROM BIOS (Read-Only Memory Basic Input/ Output System) is the key. We have to reverse engineer it without violating any copyrights. We should be able to get compatible versions of MS-DOS (Microsoft Disk Operating System) and BASIC from Microsoft."

(The BIOS was part of the operating system software. In addition to handling the input and output operations, it also managed the start-up process when power was turned on. IBM put this part of the system in ROM so that it couldn't be changed or erased.)

Sevin points out, "You need to get advice from an attorney on the ROM and follow it religiously. You can't screw that up. I'll get you the name of a good lawyer."

After about an hour, Sevin and Rosen get up to leave. Jim and I walk them out to get a taxi. They like what they've heard, but they need to do more checking before they can commit.

As their cab pulls away, Jim and I look at each other. We both feel the excitement of a race about to begin.

After the meeting we met back at Jim's house to discuss our next move. We felt pretty good about the meeting, but our recent experience taught us to not count on Sevin and Rosen. We decided to set up a meeting with Lovett, Mitchell, and Webb, an investment firm based in Houston. We also decided to flesh out the business plan to include more detail about manufacturing, cost estimates, marketing, and distribution. We were pretty sure any other investor would want to see that kind of detail.

The following Monday, Rosen called me to say he and Sevin were still positive on the portable idea and were continuing to investigate it. I was disappointed they weren't ready to invest, but encouraged to hear that they were still interested. Meanwhile, Jim and I continued to work on different parts of the business plan so we could go into more detail the next time we had an investor meeting.

The next morning, I started focusing on compatibility issues. As I thought about the way the software worked and how it dealt with the different parts of the computer, I made a list of the potential sources of incompatibilities. They included: the floppy controller, display controller, printer interface, async communications interface, keyboard, sound/speaker, memory controller, and the ROM BIOS. When I looked at the list, I realized I had included every part of the computer! It finally hit me that *every single part* of the computer would have to be designed to function exactly like the IBM PC or software applications might not work. This is going to be harder than it first seemed, I thought.

By Friday afternoon, Jim and I had most of the additional materials for our business plan ready. We had a meeting the following Monday with Jim Callier of Lovett, Mitchell, and Webb. We made copies of the plan, which had grown from its original four pages to fifteen. I was starting to get a little concerned, since we still hadn't heard from Sevin or Rosen.

At 9:00 A.M. on Monday, February 1, Jim and I sat down in Callier's office and began to describe our product idea. I went over the key product features first, and then Jim talked about how he planned to make the product more rugged than other PCs. The meeting lasted about an hour, but we could tell that this was an area Callier didn't know much about. As we left, he said he would get back to us soon.

When we got to Jim's house, there was a message for us to call Sevin. I called him back immediately. He was going to be in Houston the next day and wanted to meet with us at the airport, since he had several meetings planned. We booked a room at the Host International Hotel and began to prepare.

At 1:00 P.M. the next day, we met with Sevin and he had some good news for us. He and Rosen wanted to invest in our company to develop a portable PC—but there was a caveat. Since he and Rosen were new at the venture capital business, they needed another firm to agree to invest along with them. He wanted us to go to San Francisco and meet with Kleiner Perkins Caufield & Byers. And he wanted us to have the meeting the next Monday. After thinking for a minute, I said, "So let me get this straight. You want to invest in us, but you won't invest unless Kleiner agrees to do so?" Sevin replied that that was correct.

That wasn't what we wanted to hear, but we were excited to have the opportunity to meet with such a well-known firm. Immediately, we went back to work adding more content to the business plan. Jim created detailed schedules for the development and manufacturing phases. I thought that we might need to present our plan to several people, so I began making slides for an overhead projector. I figured this would be a more formal presentation than those we had done so far. We planned to start with a description of the product, then discuss schedules, and finally present projections of development costs and other financials.

> *"So let me get this straight. You want to invest in us, but you won't invest unless Kleiner agrees to do so?"*

We contacted Bill and told him the good news. We also said it was time for him to get on board. The train was pulling out of the station and he needed to be on it. Bill replied that he was ready to go. His infant daughter's health issue had cleared up. He would resign the next day.

Bill gave TI a month's notice, but hoped it might terminate his employment sooner. His supervisor said he had a week to finalize his projects and hand them off. Unfortunately, that meant he would miss the meeting with Kleiner Perkins.

Jim and I flew to San Francisco on Sunday morning. We checked in at the Hyatt Regency Embarcadero, adjacent to the Embarcadero office building where Kleiner Perkins was located. In the afternoon, we walked around the area and talked about the meeting. The breeze was cold, but I had a warm feeling about where this was heading. We both felt we were ready.

At 9:00 A.M. on Monday, February 8, Jim and I walked into the reception area. A few minutes later, a boyish-looking John Doerr walked out to greet us and took us back to a meeting room. I was surprised that Doerr looked so young. I was starting to feel old at thirty-seven.

Doerr introduced us to Jim Lally and Brook Byers, two partners who would also be in the meeting. I began going through the slides using an overhead projector, but I didn't get far before Doerr interrupted me with a question. The meeting became one long series of questions, with me using my slides only to address some concerns. Much of the time was spent discussing competition and how we planned to differentiate our product from others. The meeting lasted almost three hours.

Afterward, Doerr walked us back out to the lobby. He was very positive and said he liked our idea a lot, but it would be a few days before his partners could meet to discuss an investment. Jim and I were both emotionally drained after the intense questioning. We felt we had done a good job as we headed for the airport.

On Tuesday, February 9, Bill became free of TI and began to work intensely on the marketing plan. He met with Jim and me to go over where everything stood. We all agreed we needed his marketing plan ASAP.

On Wednesday, I received a call from Sevin and Rosen. They were willing to lead the initial investment round of $1.5 million, with $750,000 coming from them, $500,000 from Kleiner Perkins, and $250,000 from a third firm, L. F. Rothschild, Unterberg, Towbin. The investors would own 55 percent of the company, the three founders 25 percent, and other key employees 20 percent. I could barely contain my excitement. Then Sevin said that while he and Rosen were happy with Jim and me, they really didn't know Bill that well. He wanted to meet with him to determine if he was capable of doing the marketing job.

The next morning, Bill and I flew to Dallas to meet with Sevin. The meeting started amicably, but soon turned heated as Sevin grilled Bill on all sorts of issues. Bill believed we should sell our portables through computer dealers and planned to set up a dealer council within four to six weeks. He also strongly emphasized the need to compete on features and quality, not price. Then Sevin fired more questions at Bill. When would he hire his first employee? How many people would he need to hire? What about the European market? Would the company quibble over warranty on abused units? What about charging for service? Who is the competition?

When Bill mentioned that Osborne was the main competition, Sevin wanted to know why he hadn't mentioned IBM first. Then he wanted to know why Apple wasn't considered major competition. I felt that Bill had handled all his questions well, but I had never seen this side of Sevin before. I was a little concerned.

Bill and I flew back to Houston that afternoon, and that night Sevin called. He said that he and Rosen didn't think that Bill could handle the marketing job. They wanted to look for someone else. I thought about this for a moment. I knew Jim and I could work with someone else, but the three of us had worked closely on starting up TI's Winchester disk business. I had a lot of confidence in Bill's ability and judgment. I finally told Sevin I was going to stay with Bill, and hoped that they could get comfortable with that decision.

Sevin tossed the issue around for a while, but eventually agreed to go along with my decision. Then he shifted to other details. He told me the three founders could decide among ourselves how to divide up our 25 percent share of the company. He also said the board of directors would consist of himself, Rosen, John Doerr, and me, and we would meet at least monthly if not more often. We agreed to meet in Dallas the following Monday to sign documents and officially start the company.

Jim, Bill, and I met on Friday to finalize a few remaining details. We decided to split our 25 percent ownership evenly. Jim and I had discussed privately whether Bill should have the same equity as us, since he had not been involved in the development of the business plan or the process of selling the idea to the investors. We quickly decided that Bill's contribution was going to be just as important to our success as ours, so he should have the same ownership percentage, 8.33 percent. We also decided to give our early employees an amount of stock based on our expectations of their contribution. In a start-up, every employee has an important role to play, and we wanted everyone to share in our success.

The last issue was finalizing each of our roles in the company. Since I had five years of experience as the general manager of three different Product-Customer Centers (PCCs) at TI, I became president and CEO, Jim vice president of engineering, and Bill vice president of marketing, the same roles we had played when we worked together previously.

The three of us flew to Dallas on Monday morning and met Sevin and Rosen at the law offices of Merlin Samples. There were a lot of documents to sign, but we were so excited the time went by quickly. That day was a national holiday and the courthouse wasn't open, so papers couldn't be filed. Gateway Technology, Inc., was officially incorporated on the following day, February 16, 1982.

With money in the bank, we hit the ground running. We rented office space in the Allied Cypress Bank building on Jones Road in northwest Houston, but it was going to take several weeks to complete the remodeling. Another part of the building had been vacated recently, and the rental agent offered to let our new company use that space until our office was completed. There hadn't been time to buy furniture, so we brought in folding chairs and tables to begin working. We had a single phone line installed with one phone on a very long cord that we all shared.

But the main issue was hiring engineers. We didn't want to get our former employer upset, so we decided not to actively recruit current

TI employees—but many did come to us. Most of the people we had worked with before we left TI were aware we were planning on starting a company, and many had contacted one or more of us to let us know they wanted to apply for a job when we got started. We also decided not to tell applicants what our product was until we offered them a job. We had been trained well by TI about the importance of keeping our new product plans secret.

In addition, we decided we shouldn't offer jobs at the same time to all the engineers we wanted to hire. We picked Steve Ullrich and Ken Roberts for our first offers, because they needed to start analyzing the IBM PC. Both Ullrich and Roberts were electrical engineers in their early thirties and had joined TI right out of college. Jim had worked with them on various electronic design projects over an eight-year period. At the time Jim left TI, they were designing a Winchester disk peripheral product to go along with TI's minicomputer products. There was a strong mutual respect and trust among them. I also knew, and thought highly of, both of them.

We were planning on waiting a few days to make our third offer to Gary Stimac, another electrical engineer who had joined TI in 1972. Stimac initially was a software programmer who had worked for me on the design of the TI intelligent terminal Model 742, one of the first products to use the 8008, Intel's initial 8-bit microprocessor. I thought he was one of the best programmers I had ever met. Recently, he had been working with Jim on TI's Winchester disk project. When Stimac heard that Ullrich and Roberts had turned in their resignations, he decided he couldn't wait any longer and turned his in as well. He hadn't received an offer from us yet, and he didn't realize he was messing up our plan to spread out our hiring, but it didn't really matter. Immediately, the new company doubled from three to six employees.

At that moment, we were six people with a huge challenge. We needed an actual IBM PC to work with. Stimac was immediately dispatched to Dallas to buy one. He returned with the PC and the technical reference manual that contained detailed specifications on creating hardware and software products to work with it. Stimac, who was assigned the task of creating the ROM BIOS for our computer, discovered the entire IBM BIOS ROM code was printed in the manual. His immediate reaction was "Great, my job's been done for me! This is going to be easy."

He couldn't have been further from the reality that lay ahead.

Over the next few weeks, our team grew rapidly with the addition of four key engineers, three of whom were hired from TI. They were: Steve Flannigan, software; John Reilly, plastics; Walt Russell, aluminum chassis; and Bill Bray, power supply. Russell was working independently at the time, but we had known him from TI. They all gathered around the IBM PC we bought and, along with Stimac, Ullrich, and Roberts, began dissecting the computer. While I watched them work, I realized that we had put together an all-star team. These men were the best in their fields over all others I had met in my career.

As soon as Flannigan came on board, he and Stimac worked with an attorney to plan a legal way to reverse engineer the BIOS ROM of the IBM PC. To be safe legally, they learned that anyone who looked at IBM's code couldn't write any of our BIOS code. Stimac had unknowingly contaminated himself by looking at the code printed in the IBM manual, so his first job became writing a specification for our BIOS. Then Flannigan took his spec and began writing our code, which left Stimac with the job of working with Microsoft on MS-DOS and BASIC. We had discovered that the version of MS-DOS Microsoft was selling to everyone at the time wasn't compatible with PC DOS, so we decided I needed to set up a meeting with Bill Gates, Microsoft's CEO, and ask him for a version of MS-DOS that was totally compatible with IBM's PC DOS. We didn't know it yet, but we had just run into a potentially fatal flaw in the product idea our new company was based on.

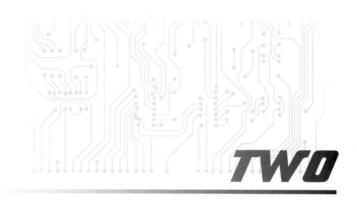

The Road Less Traveled

MARCH 19, 1982, 7:00 P.M.

I'm making my way to an aging mansion through the fog of a San Francisco night, headed to a meeting arranged by venture capitalist Ben Rosen that could make or break our fledgling company. Gateway Technology cofounder Bill Murto and I are discussing how we will handle the meeting as we ride along Market Street in the backseat of a taxi. The meeting with rising tech star Bill Gates will hold some of the first hints of what is to come in the emerging PC market.

Taking our place in line, we await our turn to have a few minutes with Gates. I believe our idea for a portable personal computer has huge potential, but we need Microsoft to make a critical change in their personal computer software to help us get our new product to market quickly.

When our turn arrives, we find Gates in one of the back rooms of the mansion. After introductions, I take a four-page business plan from my jacket pocket and lay out for Gates the important characteristics of the portable personal computer we are planning to bring to market. When I get to the software we want our PC to run, Gates leans forward in his chair.

I tell him, "Bill, we're going to build a portable personal computer, and it must be able to run IBM PC software right off the shelf. I know from my experience at TI that a start-up company will never be able to get important application software to run on its product. But if we can make our PC run software developed for the IBM PC, our customers will have access to all the important software as soon as it hits the market."

I pause, look Gates in the eye, and say, "The problem is, the MS-DOS you are currently selling us and everyone else isn't compatible with PC DOS. That may work fine for HP and DEC, but it's a nonstarter for us."

I pause again to let this sink in. "We probably can modify MS-DOS ourselves to achieve compatibility, but it would be more difficult and certainly take us longer. We need you to sell us a version of MS-DOS that is totally compatible with PC DOS."

Gates thinks about this for a long time. Finally, I ask, "Is there any reason you can't do this? Are you legally prevented from doing it?"

"Legally, I don't think it's a problem. But I'll have to ask the lawyers to be sure," Gates answers.

"So is there a relationship issue with IBM?"

"Yes," Gates says, "and it's really important for us to protect it."

"This is really important to us too, Bill. We're going to get there one way or another, but it will really help if you can do this."

Gates asks a few more questions and then leaves the room to consult with his staff. When he returns, he agrees, with reservations, to take the idea back to his programmers. His primary customer is IBM, and he has to be extremely cautious not to alienate them by giving too much support to a potential competitor. As the meeting ends, it isn't clear whether Gates is willing to sell us the software we requested.

Six weeks after the meeting, Microsoft began shipping a new version of their software to companies that had requested it, but it wasn't at all what I had asked for. Unfortunately, Microsoft didn't have the product we wanted. And after some analysis, it became clear to them that they shouldn't create the product either.

Microsoft didn't have it because it had delivered MS-DOS to IBM almost two years earlier, and although they had jointly developed the product that became PC DOS, IBM owned it. The many changes and improvements they had jointly made during those two years were not available to Microsoft's other customers. Not only that, a different group in Microsoft had made many changes to the common MS-DOS version during the same time period, so what it was shipping to their other customers was very different from what it had originally delivered to IBM.

If Microsoft had created the product we wanted, it would have caused a much greater conflict with IBM than we had perceived at our first meeting. The only way for Gates to create our product was to have his team reverse engineer its largest customer's software. Imagine IBM's reaction if Microsoft had done so and then sold it to all their competitors. It would have ended their relationship, and Microsoft would have been seen by the industry as biting the hand that fed it.

The nearest Microsoft could come to what we wanted was to find a version of each part of the software most similar to what they had shipped to IBM two years earlier. So that is what they delivered to us in early May 1982, and we took it from there.

I was extremely disappointed when I learned that Microsoft wasn't going to deliver what we had requested, but I understood its reasons.

It meant, however, that if our portable was going to run all IBM PC software, we were going to have to find—and fix—all the incompatibilities on our own.

After giving our engineers a couple of weeks to analyze the compatibility issue and estimate the impact it will have on our schedule, I call a meeting in my office. Jim, Bill, and I are under intense pressure to get our product to market quickly, so adding a major unplanned effort under these conditions is a serious problem. We discuss this with Stimac and Flannigan, our only two software engineers.

I ask, "Were you able to come up with a good estimate on the people needed and the time it'll take to get to full compatibility?"

Stimac replies first. "There are no clear-cut answers. There's just no way to know how many incompatibilities we'll find and how long it'll take to get through all of them."

Flannigan adds, "We'll need two more programmers to get started testing for incompatibilities. And we still have a lot more work to finish writing the code for the ROM BIOS."

Stimac says, "We need a strict process to test programs for incompatibilities and then find and fix those bugs. It needs to be documented, so we can effectively add more people as soon as we can find them."

We begin to realize we're facing an almost hopeless situation. To make matters worse, TI has just filed a lawsuit against us, because, we believe, we've been hiring too many of their employees. Hiring their people isn't illegal, so instead the suit claims we've stolen trade secrets. We definitely haven't stolen anything, and we're pretty sure TI doesn't even know what product we're developing. But whether it does or not, the lawsuit has to be taken seriously, which means a major drain on cash and management time. This is by far the biggest crisis our young company has faced.

To keep the meeting from degenerating further, I have the team focus on priorities. "Let's get back to basics. Do we really have a choice? Does it make sense to take our product to market with anything less than full compatibility?"

We all agree that accessing the whole IBM PC software base is fundamental to our product's success. We will not, and cannot, accept incompatibilities as unavoidable, even though we would likely be in the same position as many of our competitors. To do so would betray the core of our product philosophy.

After a while, the discussion quiets down. "So let's all be clear on this," I say. "Without full compatibility, we don't have a viable product and we don't have a viable company. We might as well shut down and give the investors the rest of their money back. We have to find good software engineers from somewhere other than TI, and we have to find a way to get the job done on time."

We had no way of knowing then that among the many companies developing IBM compatible products, we were the only one fully committed to the seemingly impossible goal of complete compatibility with no compromises. This decision would turn out to create one of Compaq's greatest differentiating characteristics and lead to the development of proprietary technology essential to our long-term success.

Columbia Data Products, also a start-up, launched its first product in June and was the first company to announce a personal computer that it claimed could run all the IBM PC's software. As it turned out, it didn't. The supposedly "fully" compatible product from Columbia had many flaws that made it highly incompatible with the IBM PC and resulted in many application programs not running correctly or at all. Columbia had rushed to be first to market, but sacrificed what we considered to be the top priority: true compatibility.

There was no way all the application software was going to work with the MS-DOS version that Microsoft had delivered to us. Even if Columbia Data Products had tried to improve their compatibility, there wasn't enough time for them, or anyone else, to find and fix all the incompatibilities that existed. But Columbia gained the media's attention by getting to market first, which encouraged many of our competitors to also rush to market without achieving full compatibility.

By late May 1982, the first prototype of our portable was coming together. John Reilly had figured out a way to get a handmade prototype of the plastic enclosure within an incredibly short time. Walt Russell had put together a prototype aluminum chassis to fit inside. Ken Roberts had been working on an idea to create a nine-inch diagonal CRT, or cathode ray tube, screen to display both high-resolution text and low-resolution graphics, something IBM required two separate screens to accomplish. Steve Ullrich had designed the circuit boards for the portable, and prototype boards were nearing completion. Bill Bray had a power-supply prototype just about ready too.

The programmers had a long way to go on our ROM BIOS and software, so they planned to use IBM parts in our first prototype. They had worked closely with Lotus, another start-up funded by Sevin and Rosen, to get an early version of their new spreadsheet application, Lotus 1-2-3. We were planning to use that software to demonstrate our portable's ability to display letters and numbers in high resolution and then, with the push of a single key, display a graph on the screen.

The design team had accomplished a lot in just three months, but a key deadline was approaching.

The National Computer Conference (NCC), one of the two largest computer conventions at the time, was being held in Houston that year. It began on Monday, June 7, 1982, and we planned to fully take advantage of the opportunity. With the help of Rosen's contacts and reputation, we scheduled meetings with a few carefully selected members of the press, several computer dealers we wanted to sell our product, and a few potential investors. Our success hinged on having a working prototype to demonstrate.

The Compaq team worked all weekend leading up to the conference, and on Sunday night almost everything was ready. But there was a stubborn, intermittent problem in the display circuitry. By 11:00 P.M., everyone had gone home except Roberts, Bray, Jim, and me. Finally, at about 2:00 A.M., we got it working. Everyone hurried home to catch a few hours of sleep before the critical meetings began.

Over the next two days, all our appointments went well. Without exception, our guests were impressed with the product's styling and its ability to run IBM PC software, especially when we pushed the F9 key and the display switched from text to graphics. The long hours of work had paid off handsomely.

We continued to look for programmers who weren't working at TI, but progress was slow. We also were progressing slowly on writing the software that Microsoft didn't provide. Stimac hadn't gotten his hands on MS-DOS and BASIC until early May, and there was a lot of coding needed to get it ready to try in our system. Flannigan was writing the ROM BIOS code on his own. To help speed things up, we outsourced some of the BIOS coding to Sail Boat Programmers, a small software company in Clear Lake southeast of Houston. Its programmers were asked to write some programs in a high-level language for Flannigan to convert into assembly language before burning into ROM.

In early August, the software team doubled in size. Charles Lee came from a small company in Dallas called DigiTech, and Paul Alito came through a recruiter. They immediately began working with Flannigan on the ROM BIOS, writing code and identifying and correcting compatibility issues.

In early September, our fifth programmer, Curt Jones, joined us from TI, and in late September, Paul Burkett came on from Sail Boat Programmers. Jones worked with Stimac on MS-DOS and Basic, while Burkett began to acquire the important software packages in an organized manner and coordinate the testing process.

With additional programmers, things began to move quickly, but then we confronted another critical problem. Some of the incompatibilities could not be corrected without Microsoft's help. When we asked Microsoft to fix those incompatibilities for which we didn't have the source code, we feared that our request would be turned down or at least acted on slowly because of their relationship with IBM. We learned, however, that Microsoft was happy to work with us and have our help in identifying problems and verifying fixes.

The Microsoft and Gateway groups began to develop a strong, mutually beneficial partnership. The collaboration between the two expanded quickly into a close working relationship, one that would deepen for years and be critical to the success of both companies.

With the positive feedback we received from the NCC meetings in early June, we began to seriously think about the need to raise additional capital. Back in April, I had hired John Gribi to fill the slot of financial controller much earlier than we had thought a controller would be needed. Now we realized we were moving so fast that we would be behind the curve if we hadn't hired him then. Gribi set up spreadsheets to manage the $1.5 million we had raised in our initial funding and was tightly controlling expenditures. At the rate we were going, he forecast that we would run out of money in early September.

Gribi, Rosen, and I spent a lot of time during July and August talking to potential investors and closing our second round of funding. On September 9, we deposited $8.5 million in the bank just as our balance dropped to zero. The company didn't skip a beat. I called it "just in time" funding, but I realized it had been a close call and that we could not afford to slow down for any reason.

In late September, Rosen set up meetings in New York with writers from the three most important business publications. We still had only one working prototype, so I flew to New York with it strapped into an airline seat next to me. Everything went smoothly until the taxi ride into Manhattan. I had the prototype sitting on the backseat next to me, and as the taxi sped along the FDR Boulevard, both the prototype and I went bouncing a foot off the seat. I asked the driver to slow down, but the bouncing continued.

The next morning, Rosen and I arrived at the office of the *Wall Street Journal* to meet with Richard Shaffer. We gathered in a small conference room and I set the prototype on the table, plugged it in, and turned on the power switch. Nothing happened. Trying hard to remain calm, I turned off the power switch, opened the computer's cover, and reseated the plug-in boards. When I turned it on again, it started normally and worked fine. I explained that the taxi ride into the city had loosened the boards. The meeting went well. Shaffer seemed interested in writing an article as soon as the product was announced. Rosen and I also met with writers from *BusinessWeek* and the *New York Times* later that day with similar results.

We had always planned on changing our company name before we launched our product. We put off addressing the issue until late summer, thinking it wouldn't take very long. The subject finally came up at our July board meeting, so we all began trying to think of a good name. It turned out to be one of the hardest things we did. Every name we came up with was already being used. We even read books on Greek Mythology to find names, but all of those ideas were taken as well.

We finally decided we needed help. Bill found a company called Name Lab in California that specialized in brand and company names. For $9,000 it would provide five good, copyright-clear names from which we could pick. At first, we were disappointed when we saw the names it provided. I thought four of them were terrible, but the fifth,

Compaq, began to grow on me after a while. We had another attorney check it for copyright conflicts and discovered that it had some. He recommended that we not use it. We went back to the drawing board, but after a week of frustration I decided we weren't going to find a name as good as Compaq, so we went forward with it and dealt with the copyright conflicts later.

In early October, our production line began to take shape. We had hired John Walker to head up the manufacturing operation. He was able to build a manufacturing team that included Bud Ronemous, Joe White, and other experienced manufacturing managers from his old company, Datapoint, and importantly, not from TI.

By the beginning of November, we had done it! We had a personal computer that ran all IBM PC software. More transportable than portable, it fit a niche that IBM wasn't competing for, at least not yet. Sales channels, marketing, advertising, manufacturing infrastructure, and financial controls were all falling into place. Less than nine months after starting the company from scratch, we were ready to go.

On November 4, 1982, in the Library Room of New York's Helmsley Palace Hotel, the company's three nervous cofounders and three other company executives, Flannigan, Walker, and Gribi, hovered around a cloth-draped object. With interest fed by rumors and carefully dropped suggestions from our well-known investor and chairman Rosen, journalists were standing shoulder to shoulder, notepads and television cameras ready, waiting to see what could be one of the most dynamic new products in the most exciting market going.

As I approached the podium, lights for the TV cameras came on and in a flash the circuits of the decades-old room immediately overloaded. All power, including to the new computer that was to be demonstrated during my presentation, was lost. Barely hesitating, I began reading my speech using the dim light coming through the windows. I announced that the company was named Compaq Computer Corporation and that our product was the Compaq Portable Personal

Computer. Within a few minutes, the lights came back on, and this time the TV cameras took turns. There were no more power outages. As I continued to talk, Jim calmly walked in front of the cloth-draped computer and prepared it for the demonstration that was to come. The remainder of the announcement went smoothly and the assembled press saw the first demonstration of a new generation of personal computers.

They seemed to be suitably impressed.

All the Compaq executives conducted press interviews for the next two hours. After the media left, I called the main Compaq office and talked to Jim's secretary, Ruth Howard. She patched the call into

From left: Rod Canion, Jim Harris, and Bill Murto with the Compaq Portable.

The Compaq team gathered outside our factory on Perry Road for a celebration of the successful launch.

speakers set up in the break room so all our employees there could hear me describe what had happened. As I told them that the product was announced and we had officially become Compaq Computer Corporation, I could hear cheers in the background. Everyone was experiencing the excitement of our success together, something I would remember for future announcements.

The next day, almost all of Compaq's 80 employees gathered in our factory on Perry Road for a real celebration. We passed out T-shirts to commemorate the event and served champagne in coffee mugs emblazoned with our new company logo. Then each of our executives who had attended the New York announcement talked about their experience and what it meant to them. The meeting quickly became an open dialogue among everyone, with my team and me answering every question. Afterward, I realized how valuable this had been for employee motivation and morale. All employees felt that they were valued members of the team and really mattered. T-shirts and inclusive meetings had just become a Compaq tradition, one that would last for over a decade.

In the days that followed, the media coverage turned out to be better than any new company could have expected. All major business publications carried articles on this new computer company from Texas. Although they didn't herald it as a breakthrough, the reports were generally quite positive and gave us incredible exposure. It is unlikely that the top executives at IBM noticed, but at least one of its sales executives did, because he would join Compaq's team within three months.

Three weeks later, Bill Gates was on the cover of *Money* magazine, and Comdex was under way in Las Vegas. The IBM PC and its supporting cast of software, peripheral, and clone companies were the show's dominant focus, stealing the limelight from the previous year's star, Apple. We set up our first booth. It was an exciting time.

The show would open Monday morning, so all afternoon Sunday, our team was busy finishing the setup of our booth. As Rosen and I walked out of the exhibit hall, we happened to go by the booth of a competitor and ran into the CEO. Rosen knew him and introduced us. I'll never forget what he said: "So, how did you get your ROM BIOS to work so well? I assume you did it like us and changed a few lines of IBM's code so it wouldn't be exactly the same."

I was shocked to hear that come out of a CEO's mouth, but I just said, "Not exactly." Later I thought to myself, *That company is in for a rough ride*. Sure enough, about a year later, IBM's lawyers knocked on its door and shut it down for violating IBM's copyrights.

With 50,000 attendees and more than 1,000 companies exhibiting, our newly introduced Compaq Portable was one in a sea of products, many of which were only recently announced as well. Many more products were yet to be announced during the show: software; graphic interfaces; Lotus 1-2-3 (already extensively tested in development on

Compaq prototype machines); a bewildering variety of new hardware products, including a large number of compatibles from major competitors, and a dozen new portable computers. Out of all the noise and turmoil, our three-week-old Compaq Portable walked away with "Best in Show." I was stunned. I believe it was mostly due to the fact that our portable was the only product at the show able to run all the software available for the IBM PC. That told me we were well ahead of other compatibles.

In an article that appeared in *Inc.* magazine after the show, I was quoted as saying, "IBM compatibility is crucial. More people are developing software for the IBM PC than for any other 16-bit computer. For that reason, we didn't sacrifice anything for IBM compatibility when we designed our system."

Indeed, our decision to do whatever it took to be able to run all the software written for the IBM PC was fundamental not only to our own success, but also to the creation of a true open industry standard for PC software as we know it today, where all PC brands run the same software.

It's amazing to me that no other company at that time exhibited the same level of commitment to total compatibility. The established computer companies were led in the wrong direction by the proprietary model that had made them successful in the past. While Microsoft could see the potential of true compatibility, it couldn't really pursue the

> *It's amazing to me that no other company at that time exhibited the same level of commitment to total compatibility.*

project without damaging its relationship with IBM. Still, Microsoft played a key role in the process through its developing partnership with Compaq.

Other PC companies that had asked Microsoft for a version of MS-DOS that could be fully compatible with PC DOS must have viewed the opportunity differently than we did. Perhaps some of them assumed that it was as compatible as possible, since it came from Microsoft. Others may have discovered some of the incompatibilities, but after realizing how hard it would be—and how long it would take—to fix all of them, they decided that "close" was good enough. Whatever their reasons, their lack of thoroughness created an incredible opportunity for us to separate ourselves from the pack and move ahead quickly. It was the first time we faced a grave, life-threatening situation, as I viewed it at that May 15 meeting, and turned it into a pivotal advantage. That pattern would be repeated several times in the next ten years.

As 1982 ended and Compaq geared up its manufacturing and distribution, the "PC Wars" of the '80s were about to begin. But it would be a long time before the final battle of the decade would decide the future of the industry.

THREE

The Best Defense
Is a Good Offense

Bill has requested an urgent meeting with Jim and me. Bill is spending almost all his time in the field trying to get computer dealers to carry the soon-to-be introduced Compaq Portable.

He says, "I think I've found something that'll really help us get dealers signed up. Almost every one of them I talk to has a story about IBM's direct sales force taking customers away from them. A few have even said that if we don't sell directly to end users, it would be a big factor in our favor. I think they all would view it that way."

After pausing to let that sink in, Bill continues, "If we can commit to the dealers that we will not sell directly to end users, I believe that it'll help us get more dealers to sign up. More important, since we need to announce several major dealers at the launch, this will get them to sign up more quickly."

I respond, "I can see the dealers' point. They're on the front lines working to get customers to buy PCs, and then once the customer decides to buy, an IBM salesman strolls in and takes the sale away."

Jim says, "How can we be sure that it won't cost us more sales than we gain? We don't know how many customers will demand to buy directly from us."

That's the dilemma. It's not at all clear which way would work best in the long term.

Bill replies, "I don't know the answer for two or three years from now, but I know today we need to sign up dealers right now. And this will help us do that."

I add, "The problem is we can't tell them we'll do it and then change in a year or two. If we commit to them that we won't sell direct, we'll have to stick by it for a long time. We'd better do some more digging."

Bill says, "I'll check with some consultants about how important they think the dealer channel will be for the next several years."

I end with "Let's meet again tomorrow. We need to get this right."

Over the next several days, the three of us talk regularly to hash out the matter. From the beginning, it's been clear that the dealer channel is now the single most important one. It's the way that most IBM PCs and Apples are reaching customers. In addition, the dealers selling IBM PCs are already trained on, and knowledgeable about, the product. Since the new Compaq machine works exactly the same way, the dealers can begin selling it immediately. Not having to provide extensive training on a new product amounts to a big cost savings for both the dealers and us.

We're aware of another big advantage from selling through dealers IBM has already authorized. A lot of small computer stores opening up are undercapitalized and likely won't be around very long. We can't afford the time or expense of problem dealers. In contrast, IBM has an intense dealer selection process and it is spending millions of dollars to ensure that its dealers are top quality and financially solid. Compaq can leverage IBM's efforts by focusing on authorized dealers.

As we meet for the fourth time on this subject, a direction finally becomes clear. I summarize, "If we succeed in getting most of the key dealers, the resulting demand should absorb all our manufacturing capacity for a long time. We can live without the additional orders we might get by selling direct as long as we're able to sign and keep the key dealers. We'll commit to our dealers that we won't sell directly to end users."

As the Compaq Portable was coming together in the fall of 1982, the company was preparing to shift from design to production. Having just closed our $8.5 million second round of venture capital in early September, we had the money we needed for the launch. No matter how exceptional the product might be on announcement day, however, the single most important question from the press and the industry would be "How are you going to sell it?" After all, products with brilliant technology and top quality can come and go as merely interesting blips in the flow of the market, if they fail to get into the right distribution channels.

There was a tremendous amount of competition for shelf space in the established sales and distribution channels. Most of the computer companies entering the market seemed to be following IBM's lead by trying to develop all the channels at once. IBM had seven different channels through which it sold its PCs, from a very strong direct sales organization to individual retail outlets. There was a lot of conflict between many of those channels that didn't seem to bother IBM, but could a small start-up handle that conflict?

In the end, we were confident we had made the right decision. The proposition we would offer dealers was truly compelling for a number of reasons: The Compaq Portable was a high-quality product at a competitive price; since it was a portable, it didn't compete directly with either IBM or Apple; it satisfied the significant demand dealers were already seeing for a portable version of the IBM PC; it didn't require much additional training for sales and service people; it didn't require stocking additional software and peripherals, since it used all the same ones as the IBM PC; and finally, unlike IBM, we wouldn't steal sales from them by selling directly to end users.

We wanted to be able to announce a strong distribution channel at the same time we launched our product, but time was very short. The two largest dealers targeted initially were Sears Business Systems Centers and ComputerLand. Sears owned all its computer stores, so signing at the corporate level meant we would automatically be in all its stores. ComputerLand, on the other hand, was a franchise operation; signing with its corporate office meant our portable would be available to franchisees, but each store decided whether or not to carry it. ComputerLand was also much larger than Sears, and every new PC maker was beating on its door to get it to carry their products.

When Bill approached the franchise's corporate officers to present to them, the acceptance process progressed slowly. He was put off, shuffled from person to person, and generally not shown much interest. ComputerLand corporate simply wasn't impressed. The Compaq Portable was placed very low in its product priority rankings, and the contract eventually offered to us had such poor terms we couldn't accept them.

Then Bill came up with a very creative alternate approach. He knew that ComputerLand had a product selection committee, made up of representatives from several of the largest individual franchisee stores. These dealers had some influence with the company's corporate officers and could recommend a new product to them. So Bill turned to the members of that committee for support in getting the Compaq Portable accepted.

He began contacting them one by one to arrange demonstrations. Committee members were based at ComputerLand's individual stores scattered around the country. Meeting with them required flying from city to city, while carrying a very valuable—but quite fragile—prototype of our 28-pound portable. Against all odds, the machine performed flawlessly in each demonstration, including one notable occasion when the prototype was positioned on top of a toilet seat in a restroom to access an electrical outlet!

Bill's demonstration of the Portable's total compatibility with the IBM Personal Computer was an utterly new and unexpected feature to committee members. He would show them the machine's quality, strength, and design features, all impressive on their own, but when he took IBM PC software out of its packaging, placed it in the Compaq Portable, and dealers saw the software perform flawlessly—they were, in a word, astonished. They had never seen anything like that before, which meant, most likely, that none of our competitors had yet achieved full compatibility. The dealers were sold, and each committee member recommended that corporate carry our portable. It was a "must have." But still, ComputerLand's corporate office remained unimpressed.

With ComputerLand taking up the bulk of Bill's time, he and I divided the dealer channel sales efforts. I worked on Sears Business Systems Centers as well as some large independent dealers. As the clock ticked down to the November 4, 1982, announcement, Sears came through, inking our first major reseller agreement.

ComputerLand Corporate somehow found out about Sears, because the very next day its executives called, saying, "We have to make a deal." They knew our product's announcement was imminent and they didn't want to be left out while Sears was included. We realized, however, that announcing a deal with ComputerLand before the contract was signed would put us in a weaker negotiating position afterward. As much as we wanted to announce a deal, we decided to stick to our basic principles and do what we thought was best for the long term. We needed a signed contract in hand before announcing ComputerLand Corporate.

When the Portable was announced in the Library Room of the New York Helmsley Palace, it was introduced with major resellers to bring it to the public. While we didn't get to include all ComputerLand stores, we did have five of its largest franchisees, along with Sears Business Systems Centers and several large independent dealers. We were able to clearly show the Compaq Portable had been solidly accepted by the dealer channel, which was a strong indication that end users would accept it as well. That was enough to encourage reporters attending the event to write favorable articles about the chances for success of our new company and new product.

Shortly after the announcement, we signed a contract with ComputerLand that included reasonable terms for both sides. Even though the product had finally been accepted by corporate, each individual franchisee still had to be visited and sold on Compaq as well as our portable. Dealers had already seen new companies come and go, so they were rightfully concerned about this one's probability of surviving and its ability to produce enough units to meet what was shaping up to be a large demand. Bill and I made a point of telling every dealer we visited about our venture capital backers and the $10 million we had already raised. It was an important part of every presentation.

To our continued amazement, each time Bill or I demonstrated the product and gave the sales pitch, by the end the dealer was asking how soon he could get delivery of five to twenty-five units. It seems customers were already asking for this product, not by name, but by capability. They wanted "the portable version of the IBM PC."

By late 1982, we were signing dealers in rapidly increasing numbers. Our decision to aggressively go after all IBM's key authorized dealers was yielding excellent results.

Many of the dealers we wanted to sign weren't part of a chain. We needed a contract with each dealer, but hadn't had the time to create one. Shortly after we announced the Compaq Portable, Mike Swavely had joined Compaq from TI to focus on developing a marketing plan, but that had to wait. Swavely was instead assigned the task of drafting an independent dealer contract as soon as possible. It didn't take him long, and, as soon as that was finished, he pitched in and helped get independent dealers signed up. Bill and I both noticed that this young marketer had a lot of potential.

In January 1983, we hired H. L. "Sparky" Sparks from IBM as our vice president of sales. Sparks had set up IBM's PC dealer network, so there was no one who knew the dealers better than he did. He quickly began to build a sales team with some of his former colleagues from IBM, and together they continued the rapid pace of signing new dealers.

During this time, I was becoming more and more convinced that there really was tremendous, pent-up demand for a portable version of the IBM Personal Computer, and the Compaq Portable looked like the answer to that demand. It was also becoming clear to me that our projections were quite underestimated.

Our initial factory was a small, leased, metal building on Perry Road, a semi-rural backroad in Houston's far northwest suburbs. It literally had cows grazing around it. Our first production estimates had been to start by building 200 computers a month and then ramp up production to 2,000 a month by the end of the year.

But based on the number of authorized dealers signing up to carry the product, along with projections of what a typical dealer could sell, it was clear that demand would surpass planned production capacity right from the beginning. Just getting demonstration units to those dealers who had already signed up would require a faster ramp than we were planning. Compaq was headed toward securing the coveted third position with all the major dealers, right behind IBM and Apple—and the product wasn't even for sale yet. The manufacturing capacity definitely needed to be much larger than we had planned.

While this was an exciting opportunity, it also carried some serious risks. Our executive team realized the good fortune of being carried by so many dealers would quickly vanish if we were unable to supply the dealers with enough computers to meet demand. Even if they liked the Compaq Portable best, dealers wouldn't hesitate to fill the demand with a competing product that was immediately available. In addition, if Compaq was somehow able to ramp up production fast enough to meet demand, there was risk of losing control of quality. There was also an increased risk of simply losing control of our inventory and cash flow and crashing into bankruptcy, as several of our competitors ended up doing.

One alternative we considered was to outsource our manufacturing to Asia, as many of our competitors were beginning to do. There were several large Japanese and Korean companies aggressively pursuing that line of business. They promised very low cost and almost unlimited quantities. That option had serious drawbacks from our viewpoint. Technology in the PC industry was moving faster than any industry before in history; in fact, it was changing almost daily. While the demand for this type of product was rapidly increasing, predicting the exact number of which product would be selling just six months in the future was impossible. The longer the supply chain for our products, the slower our ability to respond to the market's twists and turns. And we wanted to be able to react quickly to market changes.

It also was a time when revealing technology to Japan seemed risky. The Japanese business and consumer technology companies were on an extended run of expansion, which resulted in the demise of many American companies. Household names in electronics were falling by the wayside and being replaced by Sony, Panasonic, and Hitachi, among others. These companies learned from American technologies and then successfully replicated, repackaged, underpriced, and outsold them. We knew that the unique proprietary technologies we had worked so hard to create were fundamental to our long-term success. Once out of our control, they might be lost quickly to potential competitors.

We spent a lot of time in meetings working to balance the risks we would be willing to take with our desire to capitalize on the opportunity in front of us. There was no safe path available.

Gradually, our direction became clearer. We would expand manufacturing capacity as fast as possible to take advantage of our window of opportunity with dealers, but in an orderly manner and at a rate we were confident we could control. At least initially, manufacturing needed to be in the United States and under our watchful eyes so we could make changes quickly, control our lines of supply and delivery, protect our trade secrets, and maintain the highest quality standards. We did not want to lose what we believed was a once-in-a-lifetime opportunity to sit next to IBM and Apple in the dealer channel lineup.

Deciding to manufacture domestically led directly to another far-reaching decision. We knew that if we did not manufacture offshore, we could never be the low-cost supplier and therefore wouldn't be able to compete on price. Instead, we would have to be committed to competing on quality and performance. We were certain that in between the two ends of the spectrum—with products offering neither the lowest price nor the highest performance and quality—there was almost no chance for our success. Therefore, we would have to make a long-term commitment to building an upscale, high-end brand. Every product choice would have to be guided by that principle.

While it didn't take us long to make this decision, it was clearly one of the most important and would develop into one of Compaq's primary strengths in its ultimate showdown with IBM years later.

The decision to ramp up manufacturing at a rate to satisfy anticipated dealer demand meant that we had to immediately raise capital. It would require more additional investment than the total $10 million we had raised so far. To raise that much, we would have to show investors a plan that would inevitably be met with some difficult challenges.

When Compaq's CFO, John Gribi, updated his projections for 1983 with new estimates on the number of units the dealer channel would buy, it showed that our sales would be about $100 million.

We knew the number would be big, but we were shocked by this one. If we used that number in our investor presentations, who was going to believe it? On the other hand, if we reduced our projections to a more believable number, how could we justify raising enough capital to fund the ramp we needed? In the end, we decided to raise an additional $20 million and justify it with projected 1983 sales of $80 million. We figured investors probably wouldn't believe that projection either, but even if they expected sales of only $30 million or $40 million, achieving that much still meant that Compaq would be one of the key players in the PC market. I thought that position alone justified their investment.

We closed our third round of venture capital financing of $20 million in late March 1983, which gave us the money we needed to execute the upgraded production ramp planned. Whether we had the experience and systems we needed to execute the production ramp, however, remained to be seen. We didn't realize that we were attempting something never before achieved by an American company: $100 million in sales in our first year.

We didn't realize that we were attempting something never before achieved by an American company: $100 million in sales in our first year.

As Compaq's manufacturing capacity grew and monthly shipments increased, it was like pouring gasoline on a fire. When early customers carried their Compaq Portables around the country, they became walking advertisements for the instantly recognizable new product.

Then, as dealer sales increased, that led to more intense demands to speed up the process for accepting dealers who hadn't yet been authorized by Compaq. Underlying this was strong growth for the entire PC market all that year, partly driven by IBM's March announcement of a hard-drive version of its PC, the XT.

Six months later, in October 1983, we introduced our own hard-drive product, the Portable Plus. It was the first PC, portable or desktop, to have a hard drive with three-dimensional shock mounting, which was the only way at that time to achieve a truly rugged product. It was also a key differentiator for our product that otherwise would have been viewed as "me too." As with the original Compaq Portable, the Plus was met with immediate market acceptance and rapidly increasing demand.

By the end of the year, it turned out that our sales estimates for our first year were in error. Instead of the impossibly optimistic $80 million we had projected in our fund-raising solicitation—we sold $111 million worth of computers and set a record for the highest first-year sales of any company in American history.

As amazing as sales were, even more astonishing was the fact that we had somehow been able to ramp our monthly shipments from about 250 units in January to 10,000 in December, and without losing control of quality or inventory. The original factory had been designed in September 1982, before we began to realize how big demand might be. Maximum capacity was about 2,000 units per month, so we moved manufacturing to a larger building in May with a capacity of about 7,000 a month. Fortunately, we had negotiated an option to double our space in this building, so as demand continued to increase through the summer, we were able to increase our capacity a second time in October.

Increasing production was a much bigger problem than just adding more space. We had to hire and train hundreds of new workers, which would have been almost impossible to do in Houston during most years. The energy industry was experiencing a downturn in 1983, however, so a lot of people were looking for work. A local TV station carried a story about a new company that was hiring, and the next morning at 7:00 A.M. a line of applicants stretched down the block from our office. They weren't experienced in electronic assembly and testing, so the manufacturing team had to set up a rigorous training program to ensure we would still maintain high quality while adding so many new workers.

A key element of Compaq's manufacturing success was our culture. John Walker and his management team had been carefully selected for their deep experience and commitment to quality. They emphasized a family-like culture where all workers were treated with respect. These associates were given regular updates on the plan, asked for their opinions, and encouraged to speak up with ideas for improvement. There was a "can-do" attitude that permeated the organization from the top down. All employees were included in regular company meetings, where I would talk about the big picture and remind them that their jobs made a difference and they were critical to Compaq's success. It was fun to be valued members of a successful team.

Then there was the issue of increasing the flow of parts from our suppliers. An extreme shortage of some key components had developed, so Compaq's material buyers had to do more than just place orders. Wayne Collins and his team had to convince our suppliers that Compaq was going to be a winner and we should receive a disproportionate share of the available parts. This effort was helped by news of the $20 million investment Compaq received in March and by favorable coverage in the press all year. Even so, it was still somewhat of a minor miracle that Compaq was able to hire the people, buy the parts, and expand the production line enough to achieve such incredible growth, all the while maintaining excellent quality.

By late in the third quarter of 1983, we began thinking about expanding into Europe. Bill and I set up a trip to a computer conference in Hanover, Germany, to look at the competition. While there we met with Sevin and Rosen, who were then two of Compaq's four directors. We decided to look for an experienced executive in one of the major European markets to begin the expansion process. Sevin brought up the name Eckhard Pfeiffer, a German national working in Dallas for TI.

I called Jim Eckhart, a manufacturing executive who had recently joined Compaq but whom I had known for nearly thirteen years at TI. Eckhart gave Pfeiffer a good reference with only a few caveats. Then I immediately called Pfeiffer and, after a brief discussion, set up a meeting with him upon my return to Houston. Within two months, Pfeiffer had joined Compaq and began to develop a plan to enter the United Kingdom, France, and Germany.

Before the year was over, we had one more important move to make that would address our capital needs and further raise our profile. We began planning to raise capital in the public market, which meant that I would need to make presentations to Wall Street and do interviews with analysts. This was something I had never done, and I faced it with more than a little trepidation. Rosen sensed this, so he set up a session with Dorothy Sarnoff, a well-known New York speech coach and image consultant. During the one-hour session, Sarnoff pointed out that the thick eyeglasses I wore interfered with my connection to the audience. As Rosen and I were walking back to his office, we passed an optical shop that advertised contact lenses in its window. We went in and walked out an hour later without my glasses. Looking back, I realize Rosen had a real challenge preparing me for prime time. He was a great mentor.

On December 9, 1983, we made our first public stock offering. We had waited until October to begin the process for an initial public offering (IPO) on the recommendation of our investment advisors. They counseled us to wait until after our first profitable quarter to file, even though many companies in our industry were successfully completing offerings before they had any sales at all. Following their advice almost turned into a disaster.

The IPO market, which had been very hot all that year, began cooling in November. Even more threatening—and unknown to our advisors and us—IBM was about to begin showing a portable computer to large customers and dealers prior to its public introduction. If we had waited even a few weeks, the rumors alone could have eliminated any chance of our success. Without that capital, Compaq could not have sustained the growth track we were on.

Fortunately, the IPO was completed just in time. Although the company had set out to raise $100 million, in the end we raised only $66 million, but that was enough to fund even more extraordinary growth in the coming year.

The Compaq Portable's ability to run all the IBM PC's software was a key factor in the rapid and broad acceptance we achieved with IBM's authorized dealers. Compaq's record first-year sales in 1983 would not have been possible without those dealers. Just as important, capturing the shelf space of so many key dealers prevented our competition from gaining a significant dealer presence.

That's because almost all the dealers would carry only three or four major brands at one time. With IBM, Apple, and now Compaq taking up three of the spots, only one, at most, remained to be fragmented among the hundred or so remaining competitors. To lock our competitors out of that critical shelf space, all we had to do was keep our dealers happy. We stayed focused on meeting their needs.

The third dealer slot should have been owned by one of the established brands, such as HP or DEC, but their products, while providing several unique attributes, lacked most of the advantages offered by the Compaq Portable.

In terms of a baseball analogy, our product was a solid single. Being a portable that didn't compete directly with IBM or Apple made it a double. Then it stretched into a triple by using all the same software and peripherals as the IBM PC. Not selling directly to end users completed the home run. By keeping all other competitors out of the critical dealer channel, we had turned a solid single into a grand slam home run.

> *[A]ccomplishing the impossible had begun to seem like the "new normal."*

Our aggressive offense had turned out to also be the best defense, far better than we ever could have imagined. What remained was the execution of an impossibly steep production ramp that continued on into the next year. But by then, accomplishing the impossible had begun to seem like the "new normal."

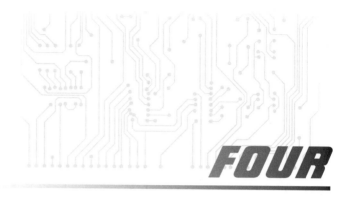

Opportunity Knocks

IN THE SPRING OF 1983, I drove to work along state highway 149, a two-lane blacktop deep in dense East Texas forest. Twenty-five miles from downtown Houston, just where the isolated highway crosses a sandy, flood-prone creek, a shiny black office building would appear suddenly out of the thick pines and my heart would race a little. It was Compaq's newly leased headquarters.

With hardly more than a hundred people, we needed only part of the modest building for our office staff, while others remained in our factory nearby. The atmosphere was electric, intense, and fast-paced. Nonstop discussions, debates, and on-the-fly decision making spilled from offices to halls to the parking garage. It had been only a few weeks since the first shipments of our portable, and it seemed to be taking off more strongly than even our most optimistic estimates had projected.

We were anticipating that IBM would soon bring out a hard-drive version of its PC, and on March 8 the announcement came about the IBM XT. We knew all our competitors would be racing to market with their own hard-drive products as quickly as they could. Our product's portability made the technology issues trickier because these small hard drives, only 5¼ inches wide, were very fragile and could not survive the bumps and shocks a portable computer would surely experience. Since a core feature of Compaq's market positioning was the exceptional ruggedness and reliability of our portable, meeting the high standards we had set was going to be a serious challenge.

MARCH 28, 1983, 9:00 A.M.

Jim, Bill, and I are well into one of our almost daily meetings. This one's in Bill's sixth-floor office with a view of the dark green forest below. We're deep in conversation about our next product, our response to the IBM XT.

Jim says, "We've come up with a solution to the hard-drive ruggedness problem, but there are some trade-offs involved. Instead of a 5¼ inch-wide hard drive, we can use the newer 3½-inch drive and shock-mount it inside a metal enclosure that is the size of a 5¼-inch drive."

I ask, "What trade-offs?"

Jim replies, "The smaller drives aren't in production yet, so we'll have to wait awhile before we can get enough units to begin shipments of our product. Also, we need to do more reliability testing to be sure it's ready for production. And they cost more."

Bill frowns. "We really can't afford to be too far behind our competitors or the dealers will start carrying some of them instead of us. And our price needs to be in line with IBM's. Isn't there any way to make the 5¼-inch drive work?"

Jim grins. "Sure, if you don't mind waiting a year while we design a bigger case."

I say, "I'm particularly concerned about a reliability problem popping up because it's so new. What do we know about the manufacturer?"

Jim answers, "The name is Rodime. It's a small company based in Scotland, but it's been building 5¼-inch drives for a while."

We continue discussing the issue for a good while. There is no straightforward solution. Running out of time, I finally say, "I don't like taking such a risky path, but I don't see any better alternative. We've got to have a hard-drive version of the portable, and there's no way to make it rugged without shock mounting a 3½-inch drive. But we have to be sure the product is reliable before we ship it, even if it means delays." On that point we all agree.

"Shock mounting" a hard drive, using rubber spacers to absorb energy when the unit is bumped, seems extreme today. The three of us had experience working with the small Winchester drives from our days at TI and knew just how fragile they were. We were also very serious about maintaining our reputation for ruggedness and reliability, because we knew that such a reputation was hard to achieve—but easy to lose.

A few weeks later, the three of us met on an issue we had not anticipated. While going through the reverse-engineering process with the new XT, our engineers found that it would not run all the IBM PC's software used in tests for our own compatibility. While it ran most of the software correctly, the fact that any of the programs had a problem was surprising.

It was hard to understand how IBM could have allowed this to happen. Adding a hard drive to its PC was a straightforward engineering task that should not have required significant changes. It seemed

IBM simply didn't place a high priority on testing the XT for backward compatibility with its PC. That meant some of its customers who upgraded from a PC to an XT would have to buy newer versions of some software because the old ones wouldn't work. Free updates didn't exist at that time.

We briefly considered whether we should duplicate the incompatibilities of the XT, but quickly concluded that didn't make sense. Customers would want to be able to use their existing software when they bought our hard-drive product. We used our compatibility technology to enable ours to run all the XT software, as well as all the PC software.

OCTOBER 25, 1983, 11:00 A.M.

With less than six months of development and not quite a year since introducing our first product, I'm standing behind a podium at Tavern on the Green in New York City's Central Park. A red cloth is covering our new computer as it sits on a pedestal next to me. After making a few introductory remarks to the assembled press and industry analysts, I lean over, pull the cloth away, and let it fall.

Revealed is a computer that initially looks identical to the original Compaq Portable. I say, "It looks the same, but the Portable Plus has significant differences. Most important among them is a built-in, 10-megabyte hard drive."

Then I hold up the new device for inspection. "Compaq has solved the problem of hard-drive fragility by going with a new 3½-inch drive and using a new packaging approach."

Step by step, I point out how the Portable Plus fits together, showing how the much smaller, 3½-inch hard drive is housed within a strong 5¼-inch aluminum cage and is triple shock-mounted.

"The drive's small size means it weighs less than its larger cousins, so there's much less energy from its motion for the shock mounting to absorb. It's virtually immune to the jarring that comes with portability."

Later I discuss compatibility. "Apparently other PC companies often confuse the compatibility issue, thinking it only means running a certain version of Microsoft's DOS, whereas most popular PC programs require a much deeper level of compatibility.

"Compaq's definition of true compatibility means being able to run all IBM PC and XT software off the shelf without modification. This capability has been a fundamental premise of our company since its beginning."

Near the end of the event, answering a reporter's question, I lay out what will become a central theme in the battle for the future of the industry:

"IBM has set a standard that all the third-party hardware and software companies have adopted. So many customers are using software and hardware based on it, the standard doesn't really belong to IBM anymore. That sounds strange, but if you step back and think about it, even if IBM didn't support it there are so many hundreds of thousands of users out there that it will continue to be the standard of the industry. If IBM comes out with a new architecture, it will have to fight on its own to set a different standard."

I get this question often during the next few years, and although my answer remains the same, my confidence in that viewpoint keeps increasing.

To most observers, the Plus seemed to be an expected and incremental move; in reality, it represented two major advancements in PC technology beyond what IBM was offering. Both advances were precursors to key strengths Compaq would continue to leverage in the years to come.

By being the first to solve the problem of ruggedness in hard drives, we further separated ourselves from the pack of clones and continued to steadily build our reputation. Not just for better products than our lower-priced competitors, but actually for better products than IBM. I began to communicate my long-term goal to Compaq's management team this way: "If a buyer can have any PC on the market, we want them to choose a Compaq over IBM. We want Compaq to be the one they desire."

After investing a tremendous amount of money and resources to make our portable totally compatible with the IBM PC, we ended up with something even more valuable and unique: We developed a proprietary technology that enabled us to make the Plus totally compatible with both the PC and XT. This was the first instance of complete backward compatibility. No other PC company had this capability.

Since the modern PC era started with the original IBM PC, the first opportunity for backward compatibility to exist in a PC came with the introduction of the XT. When some of the existing programs didn't run correctly, it didn't create a big stir. There was no noticeable criticism because the expectation hadn't been set yet. The opportunity for us to demonstrate our proprietary compatibility technology was knocking on the door, and we gladly welcomed it in. Backward compatibility, or the lack of it, would become a much bigger deal over the next few years.

When the Compaq Plus was first introduced, retail dealers didn't understand what backward compatibility meant—they had to see it in action. They had already noticed some of the existing software didn't run on the XT. When Compaq sales personnel showed them the same software running on the Plus, they were quite impressed.

They could easily have ignored the issue and let it quietly pass. Generally, the dealers liked working with us and liked selling Compaq's portables, because that made them less dependent on IBM. A lot of them began to promote this unique capability as a sales tool.

Then the press picked up on it and a few stories appeared. One quoted a Compaq dealer as saying, "Compaq is more compatible with IBM than IBM." It sounded at first like nonsense, but it was true. We were more backward compatible than IBM, and we were the only company that was.

Even if IBM noticed, it probably didn't see the news as important. IBM was the industry leader and was selling all the PCs and XTs it could build. It was the company that defined how computers were supposed to work. Tiny companies in Texas did not!

With the introduction of the Compaq Plus, we set a clearly defined paradigm for bringing new technology and innovation to the market. Each new product would need to have advancements in performance and features, clear differentiation that would justify its existence. And it would have to run all existing software.

Ironically, many of our competitors were complaining about how constraining standards were. Some said standards meant the end of innovation and the industry had to make a choice. But we knew better. The emerging industry standard would enable innovation, as long as a company understood how to innovate without creating incompatibilities. No one did this better than Compaq.

In late 1983, another pivotal point in the creation of a true industry standard occurred. Microsoft contacted us about an important matter: They wanted to know if Compaq was getting into the software business. They had noticed our dealers were selling more of our version of MS-DOS than Compaq computers. Customers were buying our software to use on other PCs.

We assured Microsoft that we had no such intentions. Then we licensed our version of MS-DOS to Microsoft to make our life simpler. Our engineers had found and fixed hundreds of incompatibilities, so our version of MS-DOS had become very different from the one Microsoft was distributing. Every time Microsoft issued a new version, we had to make so many changes that the time and cost became prohibitive. By licensing our version to them, the distribution version we and others received would already have those changes.

It appeared we were giving away the "family jewels," because all our competitors would receive a more compatible version of MS-DOS. There was a significant delay, however, between the time we sent Microsoft a new version of our DOS and when our competitors received it and integrated it into their products. Each time we introduced a new advancement in one of our computers, it was fully backward compatible on announcement day. It would take months before that version of MS-DOS found its way into a competitor's computer, so we were still able to maintain our reputation for delivering the most compatible PCs.

Licensing our software to Microsoft also established the process by which we would later spread an important attribute of the industry standard, backward compatibility, across all PC brands. Compaq was the only company with the technology to make its new products fully backward compatible. Our efforts to educate customers on the value of backward compatibility during the next three years would lead to

Compaq was the only company with the technology to make its new products fully backward compatible.

a strong expectation of it and be critical to making the industry standard more independent of IBM. Making sure that all compatible PC makers delivered backward-compatible products played an important part in creating that expectation.

The fact that we licensed our software to Microsoft has remained a secret until now.

As 1983 came to an end, the Compaq Plus was seeing high demand and winning many awards. It was also about to play a key role defending against IBM's first direct attack.

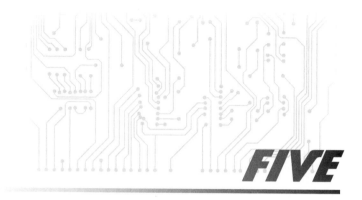

IBM Fires the First Shot

AS THE HOLIDAY SPIRIT began to fade, 1984 got off to a tense start for us. The phenomenal success of our first year in business suddenly looked unrepeatable. With production running at full speed and massive expansion plans under way, sales unexpectedly stalled. When our sales executives quizzed dealers about the sudden lack of orders, we learned that the threat hanging over us since we first started had moved from mere speculation to absolute certainty:

IBM was about to introduce a portable PC.

The dramatic success of the portable market—Compaq's market—had not escaped IBM's notice after all. We'd been given somewhat of a safe harbor in our first year because IBM hadn't yet entered that segment. As a result, we weren't viewed as a direct competitor to IBM.

That would quickly change once IBM introduced a portable of its own. Compaq would then be in direct competition with IBM, so it was reasonable to assume that its portable would take away a big chunk out of our sales just as its desktop had from other desktop makers. With IBM showing its portable to dealers and major accounts and leaking the news to the press, our dealers had stopped placing orders for our portables. They wanted to reduce their inventories while waiting to see how the IBM Portable would be accepted. Thus, the slowdown in our orders.

The press was bogging us down with coverage of IBM's rumored portable and predictions of our demise. One article paid us a backhanded compliment by saying that while we had proven to be the best of the clones, we were sure to die a slow, painful death with IBM's entry into the portable market. No doubt many potential Compaq buyers decided to wait and see.

It's difficult to grasp the hopelessness of the situation we faced. This was unlike any threat we had seen so far, and there didn't seem to be anything we could do about it. It was like fighting a ghost that was attacking us—but there was nothing for us to attack since the phantom hadn't appeared. Some of our instincts were to immediately make major cuts in production, parts orders, and factory personnel. After the initial shock, we settled into the decision-making process we had been successfully using and started considering the consequences of each alternative. Since there was nothing we could do to eliminate the competitive threat, we focused on actions we could take to deal with its potential impact.

JANUARY 16, 1984, 9:00 A.M.

My weekly staff meeting convenes in the company's main conference room. As we begin to analyze the situation, we soon become aware of just how much we've accomplished in ramping up our monthly production to 10,000 computers in December. John Walker's and Jim Eckhart's teams moved and expanded our production line twice during the year without interrupting deliveries.

Wayne Collins convinced many of our critical parts suppliers to fill orders even though there were widespread, extreme shortages.

During the explosive market growth of 1983, every computer company had forecast to capture a percentage of the market. Suppliers knew the numbers added up to a false demand higher than what was actually going to sell. We had not only followed through on our buying commitments, but also continued to raise them through the year. Our credibility was strong and delivery allocations were increasingly biased in our favor.

Collins points out, "If we cut back on our orders we'll probably not be able to increase them again any time soon. There are many other companies begging for the same parts. Our credibility with our parts suppliers would be damaged."

Bob Vieau says, "About 40 percent of the production line employees are temporary, so it wouldn't be difficult to quickly reduce their number and related production expenses."

Eckhart counters, "Yes, but cutting a large number of even temporary production workers would seriously affect morale and our momentum would be jeopardized."

Next we discuss the potential advantages an IBM portable might have over ours. We conclude it is unlikely that IBM will offer significant new capabilities in a portable that are not already present in its desktop.

Jim says, "Our portable has excellent ruggedness, expandability, and display quality. I don't see any way IBM could beat us in those areas."

Bill adds, "With a little luck, maybe it won't be as good as ours. If that's the case, once customers and dealers have a chance to check it out, I think many will still prefer Compaqs. But there's no doubt some customers will choose the IBM Portable."

Eckhart says, "But if it's in limited supply in the first few months, then I'm sure the dealers will try to sell Compaq Portables instead."

Our CFO John Gribi jumps in. "Remember, we just went public a month ago. We could face serious lawsuits from new shareholders if we stumble financially, especially so soon after raising capital. I hate to say it, but there are high expectations for our first-quarter results since we finished the year so strong."

Bill says, "If we cut back production and then the dealers' orders recover, there's no way we'd be able to fill them. The only chance we have to achieve a decent first quarter is to continue to build enough units to meet the demand that might materialize after IBM launches."

I ask, "Has anyone heard when they plan to hold the announcement?"

Sparky Sparks answers, "One dealer said he thought it'd be around the middle of February, but he wasn't certain."

As the meeting progresses, there continue to be intense feelings on all sides of the issue. Vieau and the other manufacturing executives aren't comfortable continuing to build computers we don't have orders for.

Bill disagrees. "I feel strongly that we need to be able to fill dealer demand if it develops."

Sparks, who had joined Compaq from IBM's PC group, says, "I'm sure IBM's portable will take a big bite out of our sales. We shouldn't expect orders to completely recover."

There's no consensus on this day, but the process we go through has done its job. The facts have become clear enough for me to make a decision based on the trade-off of risk versus reward. Since in this case there is no safe position, it's time for us to tighten our seat belts and keep the throttle wide open.

I summarize, "There's only one path that leads to a positive outcome, and that's if we keep building enough units to meet the dealer demand when it recovers. All other paths end up with a bad quarter financially, and I don't think it makes much difference why. We'll bet the company's future on being ready to fill dealer orders immediately after they're received. And we'll continue to order critical parts so we don't lose our place in line. However, we need to immediately revise this year's production to a lower growth rate."

Eckhart speaks up. "We have another problem. We don't have much warehouse space. So if the dealers wait too long to start ordering, we won't have a place to put the finished computers."

I ask, "Can we rent additional space somewhere?"

Vieau answers, "We can try, and in the meantime we can store some in 18-wheeler trailers."

And store them in trailers we did. Before we were through, there were almost twenty trailers parked in various locations around Houston. They wouldn't all fit in Compaq's parking lots, so we called on relationships with local companies to allow us to put one or two trailers in each of their lots. This part of the plan had the added risk of computers being stolen, but it got the job done.

Finally, after what seemed like months, on February 16, 1984, the IBM Portable, the proclaimed "Compaq-killer" product, was unveiled at a New York press conference. Whether by plan or providence, IBM had chosen the second anniversary of Compaq's founding. Even as the threat from IBM loomed, we stayed with our expansion plans. On the same day, at virtually the same hour as IBM was unveiling its portable in New York City, our management team, the mayor of Houston, and a bevy of dignitaries were standing under a brightly striped tent in a muddy field. We had shining gold-plated hard hats on our heads and shining gold-plated shovels in our hands.

As I dug my ceremonial shovelful of Texas soil, a long wagon train of Western reenactors began passing on the rural highway in front of us headed for Houston's annual rodeo. Hundreds of Old West trail

IBM fires the first shot with its Portable PC.

riders and wagons passed before us, In that moment, I was considering the extensive high-tech complex we were about to build, what IBM's announcement in New York would mean, and wondering why IBM had chosen the date of our second anniversary to launch its attack. A lot was on my mind.

After a few days of looking over the IBM Portable, the market turned out to be underwhelmed. Dealers and end users gave it average marks. It didn't offer any new technology; essentially, it was the IBM PC in a portable case. It was also heavier and slower than ours and only offered a grainy graphics monitor instead of the higher-resolution, combination display we had. Next, it didn't offer the hard-drive option of the Compaq Plus or appear to be as rugged and durable as ours. Finally, it was soon discovered that the IBM Portable didn't run a lot of existing software. Overall, it seemed like it was rushed to market, much like IBM had done with its original PC.

Still, its portable had the IBM logo attached, and it was not initially clear if its marketing power would be enough to crush us.

Within a week of IBM's announcement, a week that seemed like an eternity, our dealers began to order again. At first it was just a trickle, but in a few days it became a flood. Dealer inventories of Compaqs had run low while they waited out IBM's announcement and customer reaction. Not only was the IBM Portable a disappointment to the market, but dealers had been told they would not be getting many units for a while.

Our decision to continue building computers at full speed was paying off big time. Those trailers full of finished computers enabled us to quickly fill orders and the production lines continued to roll. Compaq Portables and Pluses were flying out the door.

Now only five weeks remained in the first quarter; it would take a super effort to manufacture and ship enough units to avoid a bad comparison financially with the previous quarter. If we failed, the predictions of Compaq's demise would still be perceived as accurate.

Things were moving so quickly there was no way to know exactly where we stood as the quarter came to a close. It had taken a monumental effort by everyone in the company to get to there. It felt like a success, but it would all come down to the final numbers.

APRIL 6, 1984, 9:30 A.M.

Gribi walks into my office and he's not smiling. He says, "I've got good news and bad news. Which do you want first?"

I hesitate. "I need some good news."

With a straight face he replies, "We beat the fourth quarter revenue. Not by much, but we beat it. And there is no bad news."

Gribi breaks out into a grin, and I jump up and hug him. "Now that's what I call good news."

When we released our first quarter results later that month, there was a stunned silence across the industry. All the media and analysts had been certain—beyond any doubt—that Compaq's high-wire act would be over once IBM entered the portable market. But IBM hadn't killed Compaq; in fact, so far it hadn't even wounded us.

When media coverage began to appear, the common explanation was that the IBM Portable was in such short supply that Compaq was merely filling in until the real IBM showed up. There was some truth to that, since Compaq was benefiting from its lack of supply.

Other forces were at work, however. IBM had given us one full year to establish ourselves and build our reputation, and we took full advantage of that time. After convincing almost all IBM's authorized dealers to sell Compaqs, we aggressively ramped up our manufacturing capacity and delivered enough computers to keep them happy. They appreciated that we had lived up to our promise of not selling directly to end users. Our dealers made a lot of money selling Compaqs.

Customers liked us because our computers did what they were supposed to do. They worked just like the IBM PC, they ran all the same software, and they didn't break when you traveled with them. We had received many reports through the year, a lot of them funny, about our computers being subjected to extreme abuse. One had been run over by a car. Another had cartwheeled down a two-story airport escalator. In every case, the Compaq survived. Stories like these appeared in the media and helped further build Compaq's reputation for ruggedness.

We received more than our share of positive publicity during that first year, partly because our portables began to show up in so many different places due to their "pick-up-and-go" quality and partly because sometimes people just like to pull for the underdog. When we had our IPO, our reported sales numbers were astounding. As a public company, we began to receive increasing coverage from the business media. By early 1984, we were on everyone's radar. We had used our one year of grace to leave all our competitors behind, except for IBM and Apple, and build an incredible reputation and following among our dealers and end users.

The secret hiding in plain sight was that the Compaq Portable wasn't a clone of the IBM PC. That's the key factor reporters and analysts were missing. By focusing on the single attribute of running all the IBM PC's software—a great quality, to be sure—they missed seeing how unique it really was. As originally envisioned, our portable was a great product at a competitive price with differentiating features and capabilities no other company—including IBM—offered. The Compaq Plus introduced in October 1983 had even more differentiation. It was the only portable on the market with a rugged hard drive, a feature that was also useful in an office if PCs were moved around. Compaq's products had a clear position in the market and a reason to exist.

When the dust finally settled after the release of Compaq's financial results for the first quarter of 1984, the media slipped back into their role of predicting trouble for us. In June 1984, when IBM announced that it had caught up with the demand for its new portable, again there was certainty in the media and among analysts that Compaq was going down.

Apparently we were following a different playbook, because we announced record financial results for the second quarter too. Storeboard, a firm that tracked PC sales through the dealer channel, reported that our Compaq Portable was outselling the IBM Portable by a factor of five to one.

By the end of the year, we announced that our sales for the year had tripled and Storeboard

The secret hiding in plain sight was that the Compaq Portable wasn't a clone of the IBM PC.

reported that Compaq was outselling the IBM Portable by seven to one. In June 1985, Storeboard reported a Compaq advantage of ten to one. In early 1986, IBM finally gave up and removed its portable from the market.

Far from IBM destroying Compaq, as so many had predicted, Compaq destroyed IBM's portable. This should have been a wakeup call to IBM that there was something different about Compaq. Instead, our success only increased IBM's resolve to return to the proprietary world it knew how to control.

It wasn't until I reflected recently on our battle with the IBM Portable that I realized how close we had come to disaster. If IBM had begun leaking information on its portable a month earlier, a skittish financial market would have shut down Compaq's late 1983 IPO. Without that capital, Compaq wouldn't have been able to continue building our portables while waiting for orders to resume.

On the other hand, if IBM had waited until later in the first quarter of 1984 to introduce its portable, our dealer order rate would probably not have recovered in time for us to save the quarter. In either case, we would have been severely injured, if not bankrupted.

IBM had unknowingly brought its portable to market on a schedule Compaq could live with.

SIX

Back at You, IBM

THROUGH THE FALL OF 1983, Compaq's management team had been weighing options to diversify our product line. A major factor in our success was deciding not to offer a product directly competitive with IBM's desktop PCs. The fit of our portables beside IBM's desktops was perfect for both dealers and customers, because the issue of their loyalty to IBM had effectively been avoided. In many ways, we believed that the safer path was to continue to be a "Portable Company" and offer a wider range of portable products.

To offer a desktop competitor to IBM would be a major change, and one with unpredictable consequences. There was risk that a desktop might be viewed as a late-to-market, "me too" product and not taken seriously. On the other hand, we had also established a reputation for ruggedness, compatibility, quality, and reliability, all traits that were important in desktop PCs.

Still, it was difficult for us to make the decision to leave the safe haven we had enjoyed.

One thing was very clear: Any desktop product we introduced would have to be significantly different from IBM's. It could start with all the key differentiating characteristics from our portables, but it would have to go well beyond those and deliver additional real, useful, and innovative features without sacrificing any software compatibility. And we would have to deliver all that at a competitive price.

The fourth quarter of 1983 had been extremely busy. We launched the Compaq Plus, settled the lawsuit with TI, and completed our IPO by early December. Valuable time was slipping away, so we made the critical decision in mid-December to start a new desktop project.

Once that decision was made, getting to market as soon as possible became a top priority. As the scheduling process began, we immediately faced an unexpected problem. We wanted to position our desktop against the XT, which meant we needed to announce it before IBM came out with a product based on Intel's new 80286 chip. Plus, we felt strongly we should not announce a new product during the summer months, when much of the press and analyst community would be on vacation. If we missed June, we would need to wait until September, adding the serious risk that IBM might announce its new product before ours. To make a late-June announcement meant we had only six months from start to finish, which seemed unreasonable for even a highly skilled, highly motivated team. After all, the Portable had been completed in record time—just eleven months.

I knew one way to speed things up was to isolate the development team as much as possible. We immediately formed a separate division led by three of our top executives: Kevin Ellington, division vice president; Mike Swavely, marketing vice president; and Gary Stimac, engineering vice president. The project was code named "Bullet" because we had to move "faster than a speeding bullet" to complete it on time. Office space at a remote location was rented for work to begin. Everything had to be done in parallel, so the marketing and manufacturing teams began their efforts simultaneously.

As 1984 began and rumors of an impending portable from IBM froze orders for Compaq Portables, what might have been shaky confidence in our plan at first soon turned into a certainty that we had made the right decision. IBM—not Compaq—had been the first to break the unspoken truce and invade our territory. Whether our desktop would be successful was still uncertain, but fear of upsetting the delicate balance was clearly no longer an issue. We were now fully committed to entering the desktop market, and in a way that leveraged the reputation we had worked so hard to achieve.

If we had not been so committed to achieving a high degree of differentiation from the IBM PC, it would have been tempting to design the Bullet product as simply a repackaging of the Compaq Portable into a desktop. That alone would have been difficult to accomplish in just six months. But we knew the likelihood of success was low if that was all we did.

Our team scoured through every aspect of the desktop PC looking for ways to make our product better without sacrificing compatibility. That was the real rub. In the last year, most of the established computer companies had introduced their own PCs that had real advantages over the IBM PC. In every case, they had sacrificed compatibility with the largest base of PC software in the world and, as a result, had failed to achieve any real success. Compaq alone had found the holy-grail technology of being able to add innovative features without sacrificing complete compatibility. That was our ace in the hole.

JANUARY 30, 1984, 10:00 A.M.

Ellington has asked me to meet with his team in the development lab at its remote Brookhollow office. They want to show me something they believe could be a very important differentiator for the

Bullet. One of our top architecture engineers, Paul Culley, explains, "We've figured out a way to use an 8086 processor instead of an 8088 and still achieve complete compatibility."

Initially, I'm skeptical. "What about the 8086 machines we've seen in the market? They've all had serious incompatibilities, haven't they?"

Culley replies, "Yes, but we've figured out why they were incompatible and have a way around the problems."

I can't help but smile as I think about what this means. "How sure are you that there won't be any surprises? We can't afford any compatibility hiccups."

"Pretty sure."

I look at Stimac. "What do you think?"

He pauses for a moment. "I believe there's some risk, but we've isolated the causes of those problems and feel sure we can fix them. It's the ones we don't know about that worry me. The key is we can get big performance advantages over the IBM PC without much additional cost."

I ask, "How big?"

"So far our tests have shown 50 percent to 100 percent improvement, but in some popular programs it's as much as three times the performance."

I let out a whistle. "Wow. Now that's worth fighting for."

We continue discussing the issues and focus on ways we can be sure of complete compatibility. It comes down to testing every application and finding a way to fix every incompatibility. When we were developing the original Compaq Portable, we wouldn't have even considered deviating so far from the middle of the road with a different microprocessor. Now we have a lot of confidence in our team's ability to achieve complete compatibility. And delivering this kind of performance advantage is the breakthrough we need to successfully enter the desktop market.

As the meeting breaks up, the decision is clear. In addition to several other unique capabilities, the Bullet will have as its lead differentiator a performance advantage of up to three times that of both the IBM PC and XT.

While the March winds and April showers came and went, the Bullet team hardly had time to notice. The thoroughness of our efforts to find areas of real improvement paid off with an impressive set of features. The Bullet would have the same dual display capability we invented for the original Portable. It would also have the same triple-shock-mounted hard-disk drives, providing ruggedness not found in any other desktop PC, along with the performance advantage of the 8086 microprocessor.

We also solved other important issues not addressed by IBM. As an example, the 10-megabyte hard drive needed a more efficient method of backing up an end user's important data than that provided by floppy disks. To back up all ten megabytes required thirty floppies and took so much time to accomplish it wasn't done very often. We addressed this issue by offering a backup option with a 10-megabyte tape cartridge, so the entire disk could be backed up on one tape in a reasonable amount of time. This kind of commonsense solution to a very real problem was exactly what we were looking for.

So were our solutions to storage expansion and plug-in board expansion. By using half-height storage devices, customers would be able to mix and match floppy disk drives, Winchester disk drives, and tape backup drives in any combination that made sense to them. By using the new 256K RAM (Random Access Memory) chips, we were able to get all 640K bytes of addressable RAM on the system board instead of taking up an expansion slot as the IBM PC and XT did. The result was four available expansion board slots in the maximum standard configuration, versus only one in the XT when comparably configured.

JUNE 28, 1984, 10:00 A.M.

About a hundred members of the press, analysts, dealers, and customers gather in a meeting room in a New York hotel. As the lights go down, we open our product launch event with a multimedia presentation that includes a laser light show drawing computer images on the screen to the beat of Irene Cara singing "Flashdance." It is the first time a computer company is using a large amount of high-tech multimedia to announce a product. It's the brainchild of Ken Price, a talented marketing manager hired from TI, and it ushers in the era of event marketing in the computer industry.

As the multimedia presentation ends, I stand in front of the audience and announce that Compaq is entering the desktop market.

"Our product is the Deskpro, and it's the fastest fully compatible PC on the market. It's also the first to offer tape backup for its hard drives."

Then I go into detail explaining all the advantages the Deskpro offers over the IBM PC and XT. To demonstrate the performance advantage, we show the displays of the Deskpro and XT side by side running various programs. The Deskpro clearly performs significantly faster than the XT.

Next, I elaborate on the storage capabilities of the Deskpro compared to the XT. Then, in an awkward bit of showmanship, I hold up thirty diskettes and begin to toss them one at a time into the audience as I explain how one tape cartridge in the new Deskpro will do the job of these thirty diskettes—and in much less time.

By the end of the presentation, I feel I have convinced the audience the Deskpro is a much better product than the PC and XT, and yet it will sell for about the same price.

The Compaq Deskpro was designed to meet user needs and perform significantly faster than the IBM PC and XT.

During the question-and-answer period, I am asked the same key question asked at the original Portable's 1982 announcement: Will Compaq's dealers carry the Deskpro, given the highly competitive battle for dealer shelf space?

I respond, "We've shown it to several key dealers and customers and it's being received very well. Really, the dealers will carry what customers demand, so in the end it's customers who'll decide about shelf space. The Deskpro was designed to meet user needs better than any other desktop product on the market, so I believe it'll get shelf space."

Indeed, Compaq's dealers carried the Deskpro, even though shelf space was about to get much tighter. Only six weeks after our launch,

Six weeks after our launch of the Deskpro, IBM announced the AT, the first PC to use Intel's new 80286 microprocessor.

IBM announced the AT, the first PC to use Intel's new 80286 micro-processor. The AT surpassed the Deskpro in performance, although not by much, but its price was much higher.

Still, IBM regained the performance leadership position with the AT, making our claim on that title short-lived. That was a negative for us, but the AT's high price point meant it would have limited real impact on Deskpro sales.

In reality, the AT announcement was positive for us because it moved the industry standard forward to the next processor level and paved the way for our future products.

The decision to rush the Deskpro to market ahead of the AT turned out to be right on target. Had we not entered the desktop market, our reputation would have remained limited to portables and we wouldn't have been able to develop a broader market leadership position. Even though the Deskpro was announced only six weeks before the AT, that gave us enough time to get our product successfully positioned against the XT and gain wide, favorable media coverage.

The Compaq Deskpro gained widely favorable press coverage.

Had the Deskpro launch missed June and occurred after the AT announcement, it would have been compared unfavorably with the AT in every area except price. It would have been perceived as a lower-priced alternative, a position we definitely did not want. In the end, our timing worked well and almost no one realized how close we had come to missing the window of opportunity.

Shortly after the AT reached the market, another positive development for Compaq emerged. The AT had major incompatibilities with the existing software base, much more so than the XT. In spite of our having repeatedly broadcast the problems created by a lack of backward compatibility, IBM still paid no attention to achieving it. Perhaps it was a matter of not knowing how to accomplish backward compatibility, since the company had never done that before. A more likely explanation was simply that IBM still held the belief that everyone would move to where it was, because it didn't need to pay attention to such details. That had always been the case. Likely, that would have continued, if we had not developed the technology to deliver backward compatibility across all our products and successfully trumpeted the advantages of doing so.

While IBM's XT had a relatively small degree of incompatibility, the AT's much higher degree became a problem noticed by almost all AT customers. IBM might have assumed that we would not be able to solve the incompatibilities of the 80286 and any issues would blow over. But it underestimated our capability and drive.

As 1984 CAME TO A CLOSE, we breathed a collective sigh of relief. It had been an incredibly challenging year, starting in the beginning of January with the sharp drop in sales of our portable ahead of the impending mid-February announcement of the IBM Portable. At about the same time, the Apple Macintosh entered the market. Then there was the extreme pressure to get the Deskpro out by the end of

June, followed by the mid-August announcement of the IBM AT and midyear price cuts on its PC and XT.

The price cuts could have been a big problem for us, and most observers predicted they would lead to our downfall. These were the same naysayers who had predicted our demise due to the IBM Portable and later the AT. This was our first real pricing challenge, so we didn't have any previous experience to guide us.

We believed that our upscale positioning would protect us somewhat and decided to make only token price cuts at first to test the strength of our positioning. The clear differentiation of the Compaq Portable, Plus, and Deskpro had the effect we hoped for and our sales continued to climb.

In spite of all these challenges, we achieved another record financial performance in 1984. Our sales tripled to $329 million and our

We completed our new factory in December 1984, just 11 months after breaking ground.

profit margins improved significantly. We had established ourselves as a major player in the desktop market and strengthened our position as the leading provider of portable computers. The new factory building we had started in February came online in December, significantly increasing our manufacturing capacity.

We also got off to a great start in the European market with the establishment of subsidiaries and product launches in the UK, France, and Germany. Eckhard Pfeiffer had established Compaq's European headquarters in Munich and recruited Joe McNally to head the UK and Bernard Maniglier to head France. The European market was lagging the American one, so our timing turned out to be excellent.

But even before 1984 was over, we had to deal with the challenge the IBM AT represented. The AT's significant lack of backward compatibility presented us with an opportunity to demonstrate our compatibility technology, but we were going to have to create a lot more differentiation than that if we wanted to continue on the path to market leadership.

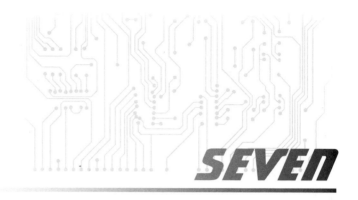

SEVEN

Patience Pays Off

A product strategy meeting is under way in the main conference room on the sixth floor at Compaq's headquarters in northwest Houston. Our product strategy team is deciding what to do in response to the IBM AT.

I say, "We obviously need to get a 286 product to market. How are we going to differentiate ours from the AT?"

The marketing team looks at each other. Then Swavely says, "We don't think we should rush to market with a 6-megahertz 286 product like the AT. We succeeded in positioning the Deskpro as the performance leader in the XT segment and in doing so added performance leadership to our brand."

He pauses for a moment, looks at me, and says, "Even though the AT is at a higher performance level than the Deskpro, it's also at the beginning of a new, more expensive price segment. We think we should wait and be the first to market with the 8-megahertz

81

286 so we can leapfrog the AT in performance. Not only that, we want to launch a portable 286-based product at the same time."

I think about this for a moment. "When will the 8-megahertz chip be available?"

Stimac responds, "It should be early next year, but there's some risk if Intel has problems."

I look at Swavely. "Aren't you concerned we'll be perceived as late getting to the 286 market?"

"If we rush to market with a 'me too' product, it'll hurt our brand image," he points out. "It's more important to get there first with the next speed increase and retake the performance lead from IBM. Plus, the Deskpro is just getting ramped up and we need to give our dealers time to absorb it. And it'll give us time to get both a desktop and portable product out together. That'll really make a strong statement about our leadership."

Everyone nods in agreement.

In making the decision to wait for a faster chip, we were sacrificing short-term sales for the long-term benefits of strengthening our upscale, high-performance brand image. The patience this decision demonstrated came from the maturity and experience level of our team. Most of our competitors rushed to get their 6-megahertz products to market and were forced to set prices well below IBM's.

Since time to market had been so crucial with the Deskpro, we had created a separate division for increasing focus and minimizing distractions. Now we were faced with developing both a desktop and a portable version of essentially the same basic product. It was an easy decision to combine the desktop and portable development teams to leverage the common features of these new products and thereby significantly improve our efficiency. Compaq was becoming a big company in some ways, but we had not lost the ability to quickly make product and organizational decisions when it made sense.

We had also developed and refined a process for ensuring a smooth transition from design to production. The strategy team would meet every week to quickly address any problem that arose. The team included representatives from every part of the company: engineering, software, marketing, sales, materials, manufacturing, production test, field service, training, and finance. The meetings were time consuming, but our experience with the Deskpro had proven its value. Swavely, who had become our key marketing executive and a strong leader, was assigned the job of leading these meetings.

We developed a close relationship with Intel by helping them find and fix compatibility bugs in

The stage was set for us to further widen our portable lead.

their first 80286 chip. Since most of Intel's customers clamored for the original version of the chip, the one that ran at the same 6-megahertz speed as IBM's, we were able to get commitments for significant deliveries of the second version with an 8-megahertz clock speed. Ours would be 33 percent faster than the AT's.

With most of the media attention focused on the desktop market after the introduction of the AT, the portable segment was out of the spotlight. IBM had not followed up with a hard-disk version of its portable, and Compaq's Portable and Plus were satisfying most of the demand for portable PCs. The stage was set for us to further widen our portable lead with a high-performance 286 model.

APRIL 30, 1985, 10:00 A.M.

We've assembled several hundred shareholders, reporters, and analysts in a hotel ballroom in Houston. They're here for the 1985

Compaq Annual Shareholders Meeting, which we've combined with new product announcements.

After conducting the annual meeting business, I say, "We've now come to the point in our program that represents one of the main reasons we're here today, and why we're so excited. In fact, we're so excited we've recruited some help to get us started on the right note..."

As I step away from the podium, the room fades to black and a strong beat vibrates through the dark. Spotlights hit the mock-up of a huge computer that is the backdrop on the stage. Then, as though performing at a concert, the Pointer Sisters, one of the hottest R&B groups, dance onto the mock computer's giant keyboard that forms the stage floor. As lasers flash, they enter in pastel-colored dresses singing their hit song "I'm So Excited." As they sing, images of our new generation of computers flash across the backdrop screen. This is not how the computer industry does things. Until now.

After they exit the stage and the applause fades, I step up to the microphone and say, "Today, Compaq is announcing not one, but two new 80286-based personal computers, the Deskpro 286 and the Portable 286."

Then I list the advantages of the Deskpro 286 over the IBM AT: a higher-speed processor, a dual-switching monitor, built-in tape backup, room for more storage devices, and more expansion-board slots.

Next I shift to the Portable 286, listing all its advantages over the AT, which are essentially the same as the Deskpro 286, but with the added advantage of rugged transportability. Then, to make a dramatic point, I show a slide listing all the competitive portable products on the market. The list is blank. There is laughter in the audience, but the point is clear: The Compaq Portable 286 is the first—and only—fully compatible, 80286-based portable personal computer on the market.

During the Q&A session, I take a question from the French audience through a satellite video connection to Paris. "Will Compaq continue to introduce new products that are compatible with the industry standard, or will you deliver innovation?"

The implication, amazingly, is that a company has to choose between compatibility and innovation, a belief still held by many in the PC industry.

I answer, "Compaq has delivered clear differentiation and innovation in every product we've introduced. Our strategy is to continue delivering innovative new features on top of compatibility with the industry standard. This is what customers want, and it's the secret to success in the PC market."

Later, I take a question from the audience watching via satellite from Munich. I paraphrase, "The question is what will Compaq do if the PC standard changes?"

I pause, shift to a more philosophical tone, and begin to address the issue of the industry standard. "Because of the millions of people who've already bought industry-standard products, the thousands of software programs already available for the industry standard, and the thousands of peripherals already available for it, the standard will not change any time soon. In fact, every day, every year, the standard keeps getting stronger.

"IBM's introduction of the AT extended and strengthened the industry standard. If IBM were to introduce a new product that isn't compatible with the standard, it would need to address an entirely different market because to not be compatible would put it at a severe disadvantage.

"In today's market, it takes more than a strong brand to be successful, as IBM found out recently." I'm referring to the poor reception its portable received. "I believe IBM understands this because

We launched the Compaq Deskpro 286 and Portable 286 at the same time.

they've continued to bring their own new products to market that do support the standard."

At the reception after the formal presentation, in response to another question, I elaborate on the issue of compatibility. "Most of the press and analysts are aware of the significant number of PC and XT programs that wouldn't run correctly on the AT. Due to the advanced compatibility technology that is proprietary to Compaq, both the Deskpro 286 and the Portable 286 run all the software available for the industry standard, which includes all software written for the IBM PC, XT, and AT.

"The issue for customers is whether or not they have to buy new versions of programs they've already purchased when they upgrade to a new 286-based PC. With the IBM AT, there are many popular programs that customers could not use, so they've got to spend additional money buying new versions just to keep doing what they are already doing. With our new products, customers will be able to use all their existing software programs. That'll save them money and make the cost of upgrading to a Compaq 286 lower than upgrading to an IBM AT."

After the meeting, there was definitely a different feeling in the audience. It was as though they were seeing Compaq as more than just a fast-growing start-up and IBM clone for the first time. In just over two years we had expanded our line from one product to seven, covering a broad range of performance and price. We had demonstrated our ability to continue to beat IBM in performance and features, and even had our own proprietary compatibility technology, which IBM didn't come close to matching.

No longer was the main question whether dealers would carry Compaq's new products. It was just assumed we had done our homework, already knew customers would want our new offerings, and were confident our dealers would happily sell them.

I had taken a slight risk by openly describing our fundamental product strategy of innovating within the industry standard, but I believed no other company had the technology to do what we were doing.

The industry standard was growing more resistant to change with every day that passed, as the number of industry-standard PCs, software programs, and peripherals continued to multiply. And now, since we had demonstrated complete backward compatibility with our 286 products, customers' desires for, and expectations of, backward compatibility would begin to grow rapidly. All the while, the PC industry standard was becoming more independent of IBM.

It was about this time that we decided to give the industry standard a shorter name. The original idea had been to create separation from IBM by not using the term "IBM compatible" anymore. That had worked pretty well, but I found myself saying "industry standard" so much that I wanted something shorter and more official sounding. So we added "architecture" to the end to make it "industry standard architecture," and then created the acronym ISA. It was quickly accepted and began showing up in the press and in conversations. To me, it made the industry standard seem even more independent of IBM.

Still, IBM had been the market leader almost from the day its first PC was announced, and it looked like it would continue to be for the foreseeable future. So it was impossible to be sure what would happen if it decided to try to change the standard. It would be a perfect example of an irresistible force (IBM) colliding with a (hopefully) immovable object, the industry standard.

The Compaq Deskpro 286 and Portable 286 were very well received. Analysts liked them and the press coverage was quite favorable. Customers buying 286-based PCs were after performance, and our new products delivered the most available at the time.

We knew our performance lead wouldn't last very long, so it was extremely important to fill demand as quickly as possible. Our manufacturing team came through by quickly ramping up production quantities for the new products. Our dealers were able to get them into the hands of waiting customers and capitalize on the initial attention the new products were receiving. That, in turn, made the dealers very happy with their increased sales and profits.

THE US PC MARKET experienced a slowdown starting in late spring of 1985. That led to another round of IBM price cuts during the summer, followed by significant cuts in clone prices as usual. We also reduced our prices somewhat, but the strength of our new 286 products helped our sales continue to grow at a reasonable rate.

Compaq finished 1985 with $503 million in sales, a 53 percent increase over 1984. That was much lower growth than our previous year, but the industry slowdown caused severe problems for most PC companies. Our 53 percent growth was viewed as a sign of strength.

In December 1985, we moved our stock listing from the Nasdaq to the New York Stock Exchange. It was a move designed to get business customers to take us more seriously and also to gain more attention from the business media. We still had to deliver good financial results, but when we did, they were more widely reported. We also had more analysts begin to cover us with regular reports.

Then, in February 1986, we announced the Compaq Portable II, a smaller and lighter version of the Portable 286. Before an audience of 3,000 employees, dealers, analysts, and media personnel packed into Jones Hall in downtown Houston, I announced, "Compaq has

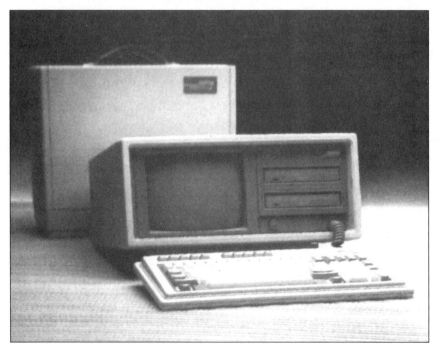

The Compaq Portable II, a smaller, lighter version of the Portable 286.

made the Fortune 500 list of the 500 largest industrial companies in America. And we did it in the shortest amount of time ever."

The celebration included a live performance by Irene Cara singing our theme song, "Flashdance," accompanied by the Houston Symphony Orchestra. More than 2,000 exuberant Compaq employees in the audience created an atmosphere of incredible energy that seemed to impress the media and analysts present. Employees left the meeting more pumped up and motivated than ever. The Compaq engine was hitting on all cylinders. Our decision to wait on a higher-performance 286 chip succeeded in strengthening our positioning and fueled a third year of record sales. Our shift to the NYSE and our entry into the rarified air of the Fortune 500 resulted in an incredible flow of positive publicity. The addition of the Portable II further strengthened our lead in the portable market. We continued to expand our dealer network around the world, and our relationships with dealers had never been better. Compaq's name was finally beginning to become a household word.

The stage was being set for the world to learn the answer to the continuing question of what would happen if IBM were to try to change the standard. But there was one more act we had to perform before that final scene could take the stage.

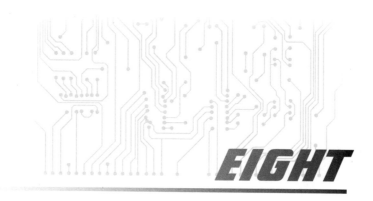

You're Going to Do What?!

Compaq's strategy team gathers in the boardroom of our new headquarters in northwest Houston to discuss a critical issue. One of our engineers has picked up on signals from Intel that seem to present us with an incredible opportunity that's also incredibly risky.

Hugh Barnes, one of Compaq's top engineers who works closely with Intel, opens the meeting. "I believe IBM isn't actively pursuing the development of a 386-based PC right now. Intel hasn't said so directly, but they seem to be looking solely at us to help make it compatible with the 286 chip."

I ask, "What's IBM up to? Why aren't they helping?"

Barnes answers, "They must've decided they aren't ready to move up to the next processor yet. And based on how much work it'll take to get a 386 PC to market, I think it'll be at least another year before they could be ready."

"Wow! That'd give us a pretty big lead. But what happens when IBM introduces their 386 PC and it's different from ours? A lot of companies have gone down because they moved out in front of IBM and then got run over."

Barnes replies, "Well, that's the big question. There certainly is risk, but we feel it's an opportunity we should take a serious look at."

I ask, "So is the 386 really compatible with the 286?"

"Not yet. Intel is trying to make it compatible, but they'll definitely need our help to get there. The last chip they sent us still has lots of problems they weren't expecting."

"How confident are you that together we can achieve 100 percent compatibility? Without that, the odds of us succeeding would be extremely low."

Barnes replies, "I believe we can get there. But it could take a while if we don't find all the bugs at this stage. Each time they have to cycle a new chip, it adds months to our schedule."

I turn to Stimac and ask, "What about the software? Do we know of any killer issues that would keep us from running all the existing software?"

Stimac replies, "We don't know of any, but there could very well be some. The 386 runs at a higher clock speed, so there'll definitely be issues. I think we can do it, but we need to look at this very thoroughly."

Then I turn to Swavely. "Mike, how do you think the market would react to the first 386 PC coming from Compaq instead of IBM?"

"I'm not sure. We need to do some research before we commit to this. We can talk to some customers we're close to and get their reaction. But we need to be very careful."

I nod in agreement. "OK. This is a great opportunity, but we have to get more information and think it through. Also, we need to find out if Intel is working as closely with any other company.

I'll call Andy Grove [president of Intel] and see if he'll commit to supporting us if we agree to launch ahead of IBM."

As the meeting breaks up, I'm already thinking about the possibilities this opportunity could create. My intuition is that we should do it, but my experience tells me to be careful. The result of getting it wrong would be disastrous.

We weren't strangers to taking risks. Early on, we developed a process for making critical decisions we simply called "The Process," but there was nothing simple about it. "The Process" evolved from a style I had developed as a manager at TI, which involved leading a small, carefully selected group of experts along a path toward consensus. The leader used the team and the consensus process to prepare him or her to make a better decision. "The Process" differed from typical consensus management because the leader retained the responsibility for making the final decision. The leader was responsible for keeping the group from getting bogged down in extraneous issues or dead-end paths. The leader also decided how long the meeting would last, and how many meetings would be held, before making the decision. Finally, accountability for the results remained with the leader.

The executive team received regular updates on the 386 at my staff meetings, which were normally held every Tuesday morning. There were also impromptu meetings called as issues arose that needed immediate attention. The excitement over what this new product could mean for our business continued to be tempered by the risks that could not be eliminated.

MARCH 12, 1986, 10:00 A.M.

Compaq's product strategy team gathers in my conference room to make a decision on the 386. I open the meeting and nod to Swavely.

"We've looked at all the issues and believe we should do it," he says. "But if we're going to do it, we need to commit now and get going because there are lots of things we'll have to line up before we can launch."

I ask, "What've you learned about customers' willingness to buy a 386 PC before IBM blesses it?"

He smiles. "There seems to be an insatiable demand for more performance. If it can significantly increase the speed of existing software like CAD packages, people will buy it. There's still the issue of IBM changing the game at a later point, but as you keep telling the press, if their 386 PC doesn't run all the current software it'll be viewed as a disadvantage."

I look at Barnes and ask, "How is Intel responding to our feedback about the compatibility problems?"

"They've really started paying attention. Early in the year, they wouldn't let us talk directly to their main processor architect, John Crawford. We had to write it all down and they'd submit it to him. Then something changed. Last month we were allowed to set up a meeting with Crawford, and now things are really moving."

"That's great. Maybe my call to Andy Grove did some good. Andy said they'd work closely with us to get the 386 to market. He didn't actually say we'd be first, but the implication was clear. I think he's starting to understand how important we are to getting his 386 chip to be compatible with the 286.

"Are there any other critical issues to consider?"

Swavely answers. "Yes. Chip availability. They made so many changes to the chip during this pass, Intel isn't willing to start enough wafers through their process to get us the number of chips we need to do the launch."

I frown. Everything seemed to be lining up, but this is a potential killer. "Is there anything we can do about that?"

"We can buy some 'risk' wafers," Barnes says. "We'd pay them to start additional wafers through their fabrication facility so that if the last version of the 386 chip they run actually works, we'd have enough chips to launch. In other words, we take the risk on the chip not working, not Intel."

"They don't sound very confident. Should we do it?"

"Absolutely. The money isn't really that much compared to what we'll be investing in everything else."

I've heard enough. "I don't want to take any chance on missing this opportunity. Let's tell them we're committed to doing it, but we want their commitment to work with us to make sure the chip is fully compatible. Even if it means some amount of delay in production."

With that decision, we dove headlong into the swirling river of fate. We were trying to achieve something no other company in an "IBM-compatible" universe had ever achieved: We were going to try to take the technology leadership position away from IBM by introducing a major advance in processor architecture ahead of them. It's hard to fully grasp just how much we were betting when we made this move. We had built an incredible reputation based on our consistent delivery of quality, performance, and business success. And we were about to risk losing it all if this move failed.

When I called Ben Rosen, our board chairman, to tell him what we had decided to do, he responded, "You're going to do what?" He was shocked at first, but after I filled him in on the details, he shared our excitement.

It was not as though we didn't have a choice. We could have easily told Intel no and waited for IBM to deliver the first 386 PC. That way we would have "played it safe" and made sure our product was fully compatible with IBM's before we took it to market. At least we thought that would be safe.

The key piece of information missing was the real reason IBM decided to delay introduction of its 386 PC. As we later found out, IBM was working on its own completely proprietary PS/2, the PC industry's equivalent of a "Death Star" that was intended to cripple or destroy every PC-compatible company on the planet. And IBM was planning on launching the new product line with a 286 processor.

If we had known about this then, it would certainly have affected our thinking, but it's not clear what the effect would have been on our decision. We might have felt even stronger about pushing ahead with a fully compatible 386 PC in order to build as strong a market position and reputation as possible before the "Death Star" became operational. But we also might have decided it was better to wait and see just what IBM would do, because the risk of IBM's new PS/2 making our 386 product obsolete was too high. Fortunately, we took the perceived "risky" path, which turned out to be the only one that could lead to long-term success.

Once the decision was made to move forward with the 386 PC introduction, every part of Compaq began to work toward making it a success. The implication of delivering the first 386 PC was we had to prepare the industry for its arrival. Whereas IBM had always been the one to run advertising and publicity campaigns to educate the market about the advantages of a new processor, now we would have to step up and take that role.

We became the focal point between Intel and Microsoft for making sure all necessary hardware and software components were ready and fully tested when needed. Leaving nothing to chance, we began to secretly work with all the important application software companies to make sure their programs worked and took full advantage of the 386 performance. It seemed to me that every

It seemed to me that every part of the company was being stretched to its limit.

part of the company was being stretched to its limit, but it was a role we had been preparing for since the beginning.

Planning for the announcement event was stepped up to a new level. I wanted to anticipate all the important questions that might be asked at the 386 announcement and thoroughly answer each one during the presentation, before they were asked. I expected there would be a high degree of skepticism by the media and analysts, so it was important to have facts, graphs, and personal testimonies integrated into the presentation. Although most of our product announcements had required less than thirty minutes, we were going to have to work hard to keep this one under a full hour.

One question particularly concerned me: "What happens when IBM introduces a 386 PC that is different from yours?" I wanted to be able to answer that one so convincingly no one would be left with any doubt. We decided to have key industry leaders speak during the presentation, strongly supporting our new product's importance and usefulness. This was intended to project the image that Compaq's 386 PC would succeed no matter what IBM did.

The industry leaders we recruited were truly the "Who's Who" of the PC industry: Bill Gates, chairman of Microsoft; Gordon Moore, chairman of Intel; Ed Esber, CEO of Ashton-Tate; and Jim Manzi, CEO of Lotus Development, creator of Lotus 1-2-3, the most successful spreadsheet application at the time.

By early September, all the stars were beginning to line up. The effort to get them there had been nothing short of monumental, but no one at Compaq was complaining. We were standing on the threshold of the future.

SEPTEMBER 9, 1986, 11:00 A.M.

We attract several hundred PC industry luminaries, media personnel, and analysts to the Palladium, New York City's popular nightclub, to announce our most important new product up to this point. They all know the product is a 386-based PC, but they're still excited because Compaq's product announcements have evolved into events that are more like parties.

The presentation opens with Rosen, who puts the significance of today's introduction in a broad context. During his remarks, he makes a bold assertion. "Today's launch is far more strategic and far more important than any prior industry announcement."

He explains why our launch is important to the personal computer industry, the software industry, end users, new users, and dealers. "This product will be the first to have the power to handle the revolutionary new software coming…that'll make the personal computer so easy to use it will become as ubiquitous on the desk as the telephone." Although many in the audience view this as hyperbole, it'll prove to be accurate in the coming years.

Rosen then introduces a peppy multimedia show set to "Headed for the Future" by Neil Diamond. The show draws on analogies to similar leaps forward in technology in other industries, including automobiles, telecommunications, and medicine. As the multimedia show ends, an oversized mock Deskpro 386 descends to the stage, literally filling the entire space.

I step to the podium. "Ladies and gentlemen, introducing the Compaq Deskpro 386 personal computer. Today, Compaq Computer Corporation is stepping to the forefront of the personal computer industry. The Compaq Deskpro 386 is the first of a new generation of industry-standard desktop workstations.

"It's by far the most advanced high-performance personal computer in the world today. Together, Compaq, Intel, and Microsoft are setting the pace for the rest of the industry with the next generation in general-purpose, high-performance workstations.

"Since the 80386 delivers compatibility with the past and the architecture needed for the future, and since it comes from Intel, there's no doubt that the 80386 will be the foundation for the next generation of the industry standard." This last statement seems obvious to many in the audience, but I want to head off any question that something else could possibly be the next-generation industry-standard processor.

Next I describe the specific features of the Deskpro 386, including performance that is two to three times that of 286 machines, increased RAM memory size and speed, increased hard-drive size and speed, and greater flexibility provided by more storage device slots and board expansion slots. I tell the audience, "We have addressed every aspect of the computer and optimized for performance."

As I finish this part of the presentation, I'm at the point where I would normally wrap up and open the floor to questions. But for this announcement, we're barely at the halfway point. Still lined up are the leaders of our key industry partners to speak about the significance of the Deskpro 386, the advantages it will provide, and their plans to support it with advances in their own products.

First up is Bill Gates, Microsoft's CEO. He talks about how excited he is about the Deskpro 386, how Intel, Microsoft, and Compaq have worked together to make it happen, and how Microsoft will support its advanced features with its operating systems. He says Microsoft has Xenix, its version of Unix, available now and that the company is working to bring an advanced version of DOS to market soon.

Next comes Intel's Gordon Moore. He tells how Intel has worked with Microsoft and Compaq to bring the 80386 processor to market and to achieve full compatibility with its previous generations of processors. He says Intel is capable of delivering a million 386 chips in 1987, meaning there'll be plenty of them available to meet Compaq's, and the industry's, demand.

The Compaq Deskpro 386 is the first of a new generation of industry-standard desktop workstations.

The two final speakers are Ed Asber of software pioneer Ashton-Tate and Jim Manzi of Lotus Development. Both talk about how their companies will take advantage of the capabilities of the 386 with their own applications, and how they're fully supporting the Deskpro 386.

I return to the podium and tell the audience that Compaq has been working with several large PC customers to test the Deskpro

386 in real user applications. I read quotes from three of them that are effusive in their praise of the Deskpro 386 and the value it will bring to them. One quote, from the Director of Worldwide Distribution at Sisters of Charity Hospital Corporation, says, "The thing we liked best about this machine is that we dropped all the boards we use in it, we dropped all of our software in it, and they worked right from the start...[New powerful machines] are only good when they're compatible."

If IBM doesn't pay attention to anything else at our announcement, it should at least listen to this customer. He has succinctly stated the power and importance of the industry standard.

I begin to summarize. "Today, we're ushering in nothing less than the third generation of the personal computer revolution. Nine years ago, the original Apple II helped start the personal computer revolution. Five years ago, the introduction of the IBM PC was the catalyst for the first true computer industry standard. Today, the Deskpro 386 is bringing us the third generation. For the first time, true 32-bit computing power, within the industry standard, is available in a personal computer. Intel's microprocessor family is delivering the next generation in performance and architecture. Microsoft's operating systems are expanding the capabilities of the industry standard. And the applications delivered by Lotus, Ashton-Tate, and others will bring the benefits of the next generation directly to end users."

Just as the audience might think I'm about to finish, I go into something they've never heard before nor expected to hear now: a ten-minute analysis of the potential responses by our major competitor, IBM. "Now I'd like to address a major question raised by today's announcement. Why has Compaq decided to move out ahead of IBM to the next generation of the industry standard? The answer is that the 80386 is of critical value right now to both the industry and the user community...It's unreasonable to expect the industry and users to wait for IBM, especially if the wait is likely to be a year or more. We have evaluated all the risks and the arguments against moving ahead and have concluded that none of them are valid.

"Let me explain why. First, it's important to understand that Compaq by itself isn't trying to set a new standard. Standards are set by users, and the industry standard we have today is in the hands of the user community. The foundation for the next generation of the industry standard is the next member in Intel's compatible microprocessor family, the 80386. The Compaq Deskpro 386 is simply the vehicle for delivering this important breakthrough to users in the best and safest manner possible through full compatibility with the industry standard. Now, even though this breakthrough is important and useful, and even though Compaq is delivering it safely, what might IBM have to say about all this?"

In the first part of my analysis, I address every component of the Deskpro 386 system and show how we have optimized each area so there is little or no room for IBM to improve upon them. "Even if IBM does introduce some unexpected new capability, expansion bus slots will allow for that capability to be added to the Deskpro 386. So there are really no significant user needs that aren't being met by the Deskpro 386."

And finally, I ask the fundamental question—the one I know everyone wants to hear: "What if IBM goes beyond industry-standard 8- and 16-bit slots and offers a new 32-bit slot?" My answer has been thoughtfully prepared and reflects the knowledge gained in developing the first 80386-based PC.

"The only thing a 32-bit slot offers is more speed. We've looked very carefully at all the peripheral areas and concluded that the only one that needs the speed of a 32-bit bus is the RAM memory. The

Deskpro 386 does provide a 32-bit bus for its memory, and therefore has the maximum speed.

"When you add all this up, I believe it's clear that a 32-bit port is not needed on this class of product, and therefore it would not have a significant impact on the 80386 desktop workstation market.

"But whatever IBM does, a key factor is when they do it. If six months or more goes by, the 80386 will gain significant momentum without IBM. A lot of users will become satisfied with the advantages of the industry-standard 80386 product. By then, I believe IBM would have a difficult time convincing the majority of the industry to follow it down a proprietary path. The burden would be on IBM to prove its proprietary offering is useful and significant enough to cause everyone to give up the very real benefits of the open environment of the industry standard."

I pause briefly to let my answer sink in. I feel I've distilled the message down to a few straightforward sentences that are easy to understand. Even though the tone of my answer sounds like I have some knowledge of what IBM is going to do, in fact I don't. I'm simply trying to head off any possibility that a reporter or analyst might form a negative opinion about the Deskpro 386's chance for success. Subsequent events will indicate that IBM did not pay attention to what I said, although hindsight will eventually show it should have.

I close with a prediction of my own.

"So far, the benefits of the personal computer revolution have been dramatic. Personal computers lead to better, faster decision making. They allow office workers and professionals to do their jobs more quickly and accurately than was ever thought possible. And they give the companies that invest in them a significant competitive advantage. The 80386 generation will bring even more potent benefits than all that's come before. To those who were involved at the start of this revolution a decade ago, at times it must seem as if things are starting to slow down. Then along comes an advance like the Deskpro 386 and you realize that truly, we've only just begun."

The Deskpro 386 event was a high point for Compaq. Every part of the company had contributed to its success, and many dedicated people had worked long hours in the summer months leading up to it. Afterward, there was little time to rest and reflect, however. The job of convincing the market that the Deskpro 386 wouldn't be instantly obsolete when IBM introduced its 386 product had just started. We knew there would be many skeptics, so we planned an intensive publicity and advertising campaign for the weeks immediately following the announcement.

Two days after the launch, we ran an eight-page pull-out advertisement in the first section of the *Wall Street Journal*. The cost was astronomical, but it was part of the price of moving out in front of IBM. Every page contained a bold headline and addressed a key point. The advertisement reflected the same thoroughness we had exhibited in the announcement.

Initial media coverage in the major business publications was generally complimentary of the Deskpro 386, but many contained quotes from analysts, who pointed out the uncertainty of what IBM's response might be and how it might affect Compaq.

My focus during the announcement on why that wouldn't be a problem for us had gone a long way toward softening the analysts' concerns, but they were far from eliminated. The one key point that did sink in was that likely, it would be six to twelve months before IBM would respond. They mostly agreed that our product would see strong demand during the interim.

The media coverage in the trade publications was more effusive with praise of the Deskpro 386. Every article recommended the product for applications that needed more power than the current 286-based products could deliver. Some of the articles went into a detailed analysis of applications where the 386 would make sense.

Media coverage aside, our customers and dealers went crazy over the Deskpro 386. We were already in production at the time of the announcement, but for a while afterward were limited by the number of 80386 chips available from Intel. The supply that was available was due to the "risk wafers" we had paid Intel to start. Our sales team worked with our dealers to make sure that as many 386 machines as possible went to companies evaluating the product. We wanted to be certain that the demand would be there when Intel's chip supply ramped up in the first quarter of 1987.

The immediate success of the Deskpro 386 was partly due to its ability to run all the existing PC software, plug-in boards, and peripherals. We had established a reputation with the Deskpro 286 for being able to do that, and thus had begun the process of educating the market on the value of backward compatibility. The Deskpro 386 raised awareness of the value of backward compatibility by an order of magnitude because it was such a vivid demonstration of its power. We proved that a company other than IBM could successfully introduce a new, advanced microprocessor as long as the product maintained complete backward compatibility. As a result, many people understood for the first time that it really was an industry standard, and not just an IBM standard. IBM no longer had to be present for the party to continue.

With the aid of a few thousand Deskpro 386s delivered in the fourth quarter of 1986, we finished the year with $625 million in sales, a 25 percent increase over 1985. In most industries, 25 percent annual revenue growth would be spectacular. But for the personal computer industry, it was just mediocre. Overall, PC demand had been weakening most of the year and it was affecting all the PC companies, including IBM.

Rosen had predicted during our 386 announcement event that the Deskpro 386 would usher in renewed growth and vibrancy to the industry. Whether it was the 386, or whether his timing just happened to be right, PCs did see accelerated growth in 1987. For us, there was no doubt the Deskpro 386 accelerated our growth.

The other thing that helped get 1987 off to a fast start was our introduction of the Portable III on February 17. Most of our portable competition was coming from laptops at this point, and we were being criticized for not entering that segment. Our strategy had been to continually decrease the size of our portables, but to make sure

The Compaq Portable III met our goal of an even smaller yet fully-functioning portable.

they remained "full-function." The Portable III filled that role well and helped us continue in our portable leadership during 1987.

Several of Compaq's executives and I mounted an intense campaign of speeches and interviews during the latter part of 1986 and the first part of 1987 to overcome any resistance to us getting out ahead of IBM. In addition, our major advertising associated with the launch was followed up by an overall higher level of advertising throughout 1987.

Meanwhile, another major event affected us during 1987. In early April, IBM announced its PS/2, the long anticipated and feared "Death Star" aimed at eliminating the clones, among them Compaq.

NINE

The "Death Star" Arrives

I have called a meeting in the small conference room by my office to be briefed on details of the new IBM computer just introduced in New York and to discuss the implications for Compaq. In attendance are Mike Swavely, marketing and sales vice president; Gary Stimac, systems engineering vice president; and Hugh Barnes, engineering vice president, along with several others.

Swavely summarizes the information. "This is a completely new computer. It's not even called a personal computer. It's called the 'Personal System 2,' or PS/2. It's got a new expansion bus they call 'Micro Channel.' Apparently it's totally proprietary. It won't accept any of the industry-standard plug-in boards. And they [PS/2s] come with 3½-inch floppies, so none of the existing industry-standard software can be installed on them."

I contemplate this for a moment. "What's the microprocessor?"

Swavely replies, "Initially it's an 80286. They also announced a 386 version, but that won't ship until late summer."

A slight smile crosses my face. "Well, that's some good news. It's going to be hard to convince people to give up access to all their current software and peripherals for no increase in performance over a 286. And late summer for their 386 means we'll have almost a year to get the Deskpro 386 fully accepted."

Swavely says, "Here's more good news, I think. They announced they're going to take the PC and AT out of production. So people who want to run their existing industry-standard software and peripherals will have to buy Compaqs."

I reply, "Wow. They're really playing hardball. They're trying to force a quick transition to their new architecture. That means we'll need to respond pretty quickly with a clear statement about our direction."

Everyone thinks about this for a minute. Then I say, almost to myself, "They've really done it. They're really throwing down the gauntlet this time."

"OK then," I say, raising my voice. "We're going to have to really dig in and understand every aspect of this. I think our sales should be OK for at least the next several months as customers digest all this.

"But we need to make good use of that time to plan our response. As soon as we can get a complete comparison on prices, let's meet and decide if we need to take any near-term action. We also need a complete analysis of all the technical aspects of their machine. We need to be absolutely sure we know what we're dealing with."

As the meeting breaks up, we're almost in a state of shock. We were aware IBM was going to make an important announcement, but we thought it was going to be about its version of a 386 PC. There'd been some tension building as we worried about how much IBM would change its 386 PC and whether it had found some way to make our Deskpro 386 obsolete. But this is a whole new ball game. It's now clear to us that IBM is trying to make all of our computers obsolete, not just the 386.

During the next few days, there were many meetings throughout Compaq to understand all the implications of the PS/2 announcement and to figure out appropriate strategies. Our engineers bought one of the first PS/2 computers and immediately began to tear into it. We also began to study the computer's hardware and software manuals. Stimac contacted representatives at Microsoft to see what they knew. Barnes contacted employees at Intel to find out if they knew anything they

The IBM Personal System 2 (PS/2) has a new expansion bus called "Micro Channel."

could tell us. Others contacted software developers to determine if they already had applications ready to run on the PS/2, and also companies that sold plug-in boards to find out where they stood.

It quickly became apparent that IBM had done a very good job of keeping details of the new product a secret from the rest of the industry. As far as our inquiries could determine, neither Microsoft nor Intel knew everything about the PS/2 before the announcement.

There were no third-party application programs or plug-in boards initially available or about to be available. If a company wanted to develop a plug-in board for the PS/2, it would first have to buy a license from IBM to use the Micro Channel. IBM was clearly very serious about maintaining proprietary control over its PS/2 and making money from everything that went into it.

In January, IBM had begun sending strong signals of imminent products that would be much more proprietary than its existing PCs. Early in the year, IBM briefed a number of its major customers about products that couldn't be cloned. Some were told IBM's next operating system would not run on computers from other companies. IBM had implied its new products would be compatible with its existing products, but would offer advances that could not be legally cloned.

Based on this information, we believed IBM would add a proprietary 32-bit bus to the existing industry-standard 8- and 16-bit buses. This approach made sense to us. But the Micro Channel bus had absolutely no compatibility with the industry-standard bus. That move did not make sense.

We had several intense meetings over the next few days. We realized there was no way to convince the media and analyst communities that IBM had made a mistake.

We decided we needed to convince a significant number of customers to think before they blindly followed IBM down the PS/2 path. To accomplish this, we decided to take a strong position on the negatives associated with the PS/2's lack of backward compatibility with the existing base of software and plug-in boards. Even if the press didn't agree with our position, we wanted it to be in print where customers would see it.

We realized we had to be careful not to paint ourselves into a corner and damage our reputation if the PS/2 did eventually become the new industry standard. And we had to be very sure of our facts when we publicly criticized any aspect of the PS/2.

The approach we settled on was to carefully and specifically point out the disadvantages a customer would face when moving to the PS/2. At the same time, we would make it clear that Compaq would always listen to the demands of the market and provide customers with products and features they desired, including PS/2 compatibility. The key was to handle our response objectively and accurately, so that we would not be viewed as defensive or scared.

We were also looking for a way to draw a parallel that would weaken the widespread perception that whatever IBM did was right. The analogy we chose was New Coke, a well-known failure of a market leader to move customers in a different direction. The idea was not to say the PS/2 would fail but to point out it was not a given the new computer would succeed to the same degree that the original IBM PC had.

By mid-April, along with Swavely and others, I began giving speeches and conducting interviews articulating Compaq's position. We quickly succeeded in getting our message widely communicated. The media loved a conflict, and we became the point man for the opposition in this one. There were so many articles being written that I asked our market research group to send me a report of all the different viewpoints. The following comments occurred in May and were included in the report that I received in early June.

For the most part, our message was understood and well received, but a significant number of people missed part of the message. Adding the analogy to New Coke opened up a hornet's nest.

Wall Street's analysts seemed to understand the balanced message best. They followed our logical argument, which established the PS/2's incompatibility with the existing standard as a negative. But they also understood we were saying we would follow the PS/2 if enough customers demanded it.

Michael Davis of investment advisor Lovett Mitchell Webb & Garrison wrote, "Most important was Compaq's stated continued commitment to being responsive to customer needs. Over the near term, Compaq will accomplish this by continuing to provide existing and new products that are compatible with the industry standard. Compaq will only introduce products that are compatible with IBM PS/2 when the marketplace demands such products. As long as Compaq maintains this attitude, we see no near-term limits to its growth."

Michele Preston of Salomon Brothers wrote, "In its most aggressive stance ever, Compaq is promoting the importance of compatibility with industry standards and providing products that provide real—not perceived—user benefits. This will strengthen Compaq's franchise in business accounts and the dealer channel... The company has been meeting with customers, the press, and analysts to make this position clear."

One who missed part of the message was Peter Labe of Drexel Burnham Lambert when he wrote, "Canion's recent speech is both brilliant and articulate. It is as persuasive a case as we have seen for a two-standard market—the IBM standard (as personified by PS/2) and a *de facto* standard of all the existing stuff of which Compaq is the most ardent exponent. We believe there is great risk this time that Mr. Canion might not walk on water. He is right technically, no doubt—but in actual practice, we will bet on IBM. In the history of the compatible industry, where there is a history since 1969 in other products, it has been conclusively proven that there is no such thing as 'almost compatible'.... The fact of the matter is that IBM has regained the initiative and leadership position."

Not surprisingly, executives from some large companies completely toed the IBM line. In response to a query from a Compaq sales representative, a senior executive of ARCO told us, "You folks are making some very big assumptions....IBM sets the standards, IBM drives the standards...and if IBM wants to it can lock out all PC vendors."

There was also relatively strong support in the PC industry press for the infallibility of IBM. A *PC Week* editorial on May 5, 1987, stated, "Users are hearing a lot of rhetoric in the aftermath of IBM's PS/2 announcement....Rod Canion publicly compared the PS/2 to Coca-Cola's well-publicized marketing disaster, 'New Coke'...the implication is that PS/2 is a radical and unnecessary departure from users' current installations. The view...is simplistic. Worse, it creates unnecessary anxiety for users who feel pressured to choose between allegedly irreconcilable architectures." He was partly correct. We were trying to get customers to think things through and not just blindly follow IBM.

Most disappointing to me was the reaction of leading industry consultants. Stewart Alsop of the *PC Letter* wrote, "Taking Compaq's statements at face value indicates that it will stick with what it has and take a different route than IBM. That particular strategy is an extremely dangerous one, if you look at the history of the computer industry: the only companies—without exception—that have prospered in the long term are ones that have either stuck with what IBM has defined as the standard (like Amdahl) or ones that have departed completely from the standard (like Digital Equipment and Apple)."

Aaron Goldberg of International Data Corp., a market research firm, wrote, "Rod's approach is obstinate...outlook for Compaq is trouble if they keep thumbing their nose at IBM....Verbal commitment to the new standards is important....Ostriches need not apply...."

Doug Cayne of Gartner Group wrote, "Compaq publicly denying the need for new architecture elements...privately working hard to imitate them...biggest risks are pricing and arrogance."

I decided the analysts didn't understand the risk associated with following IBM to the PS/2. In spite of the confusion and mixed reactions, we stayed with the same message. It was getting through to customers and seemed to be having the desired effect, since our sales continued to grow through the summer.

During all this, we met many times to discuss how we should deal with the PS/2 and Micro Channel from a product development standpoint. Our engineers concluded there were no real advantages associated with the Micro Channel. They reported that each of Compaq's industry-standard products outperformed the most comparable PS/2 product.

Adding this information to our speeches and interviews strengthened our arguments and led us to decide we had to focus part of our engineering team on finding more innovations that would continue to give our industry-standard products performance advantages over IBM's PS/2. We believed the tide could turn quickly if new PS/2 products were able to outperform ours. But we could not escape the possibility IBM's brand and marketing power might be strong enough to actually pull it off—even if we outperformed it.

We were sure many of our competitors were rushing to develop new products compatible with the PS/2, even though they had to sign a license with IBM and agree to pay very significant royalties. If the industry moved to a new standard, and if our PS/2-compatible products were behind our competitors, Compaq would almost certainly lose its leadership position. So we decided we would have to carve out a significant part of our product development resources and get started on reverse engineering the PS/2. This was a very difficult and particularly distasteful decision for me, but we were pragmatic enough to realize that waiting too long to get started could lead to disaster.

Even before our official decision had been made, a small group of engineers had started the process. After dismantling an early

PS/2 to understand exactly what it was, they began to plan an approach.

We discovered a lot more resources would be necessary to get the job done this time. The same people who had reverse engineered the original IBM PC were still around, but IBM had taken more time developing the PS/2 and it contained several special customized chips called "ASICs," or application-specific integrated circuits, which would be very challenging to reverse engineer.

We were stunned by an estimate of the people and money it would take to complete the process and design a product we could take to market. The bottom line: We were going to have to essentially split our engineering resources down the middle, half working on PS/2-compatible products and half developing industry-standard products that would enable us to stay ahead of IBM in performance. Two years earlier, this would not have worked. But we had steadily expanded our engineering and software teams and would be able to do a reasonable job of both, at least for a while.

It didn't take long for this decision to leak to the industry. From the beginning, we had stressed to employees the importance of keeping product plans secret, but there was just too much attention being paid to what we might or might not do for something this significant not to get out.

We were stunned by an estimate of the people and money it would take to complete the process and design a product we could take to market.

Included in the report I received in early June was an excerpt from an article that appeared in the June 2, 1987, edition of *PC Week*: "Compaq hedges its bets…straddles the fence on issue of PS/2. Bolstered by Compaq's burgeoning sales, Mr. Canion has been uncharacteristically strident in his criticism of the PS/2 and has eagerly assumed the mantle of standard-bearer for PC compatibility. Although he speaks from a position of strength, Mr. Canion's anti-PS/2 remarks startled many industry observers. The contrast between Mr. Canion's avowal and his design team's feverish quest [to duplicate the PS/2] illustrates the ambivalent relationship of Compaq to IBM and the parallel strategies the company must pursue to maintain its leadership position in the industry."

I was disappointed that there had been a leak, but glad the article had gotten it right. I thought it might actually have a positive effect on those who had missed my comments about following the PS/2 if enough customers demanded it.

I felt strongly we needed to continue to introduce products that clearly outperformed the comparable PS/2s from IBM, thereby supporting our position that there was no real performance advantage with the Micro Channel. And I really wanted to make the point there was more performance to be gained from our technology advancements than from the Micro Channel. My chance came with the product we announced in September 1987.

In the process of working with Intel to get the first 386 to market in 1986, we had mainly focused on making sure the chip was totally compatible with the 286. During that process, our engineers learned the performance of the 386 was severely limited because it couldn't access the memory and Input/Output buses in parallel. In other words, if an operation with a peripheral was in progress, the processor couldn't access memory until the I/O bus was free.

Paul Culley, our leading hardware architect, had begun working on a way to fix this bottleneck as soon as the Deskpro 386 had gone into production. He came up with a creative solution that was being implemented in the next product, the Deskpro 386/20, with the "/20" signifying a 20-megahertz version of the 386 chip.

A memory cache controller took over both the memory and peripheral buses and accessed them in parallel. In addition, while Intel had a good math coprocessor for the 80286 called the 80287, it didn't have one ready for the 386 when it was going to market. We had used the 80287 as the math coprocessor for the Deskpro 386, but it was far from the optimal solution. Since Intel hadn't caught up with a viable 80387 chip, we chose to work with the chip-design company Weitek to use its math coprocessor chip in the Deskpro 386/20.

The result of these two innovations was that the Deskpro 386/20 would run more than 50 percent faster than other 20-megahertz 386 machines. And since the PS/2 Model 80 used a 16-megahertz 386, the Deskpro 386/20 would run almost 100 percent faster. That was the kind of clear performance advantage I wanted to be able to demonstrate.

On September 29, 1987, we announced two major new products, the Deskpro 386/20 and the Portable 386, which also used the 20-megahertz 386. In order to highlight the unique technology we had invented, we created the name "Flex Architecture" to refer to the combination of features that resulted in such amazing performance advantages. The announcement touted the performance of both products, and the demonstrations held afterward clearly showed them as cutting-edge.

In what could have easily been a ho-hum announcement of a clock speed upgrade, we succeeded in making a very strong statement about our leadership in PC technology. Our timing was nearly perfect, since the first PS/2 386 machine had just started shipping. We were able to back up our strong performance claims with simple demonstrations comparing the Deskpro 386/20 and Portable 386 with the IBM PS/2 Model 80. The performance differences were so clear they couldn't be ignored or explained away. We really were far ahead of IBM in delivering high performance, and we did it with a machine that ran all the existing software and all the existing plug-in boards as well.

Most of the press and some of the analysts got the message we were delivering. In fact, the coverage of our two new machines was the most positive we had ever received. In addition to performance, two other aspects really seemed to make an impression. One was our name for the "secret sauce" that made our machines so much faster. By calling it "Flex Architecture" and referring to it over and over by that name, it began to take on an air of significance similar to IBM's Micro Channel. Although IBM had continued to tout the performance advantages of Micro Channel that were "coming soon," we demonstrated that our performance advantages were "here and now."

The other thing that made a strong impression was that we were able to put a machine with all this performance into a portable package that weighed just over 20 pounds and was much smaller than the original Compaq Portable. As we stated in our advertisements, it was "pound for pound the most powerful computer on the planet." This was one more example of Compaq doing something no other company had ever done. The positive impact on our brand image and reputation was simply beyond measure. Compaq was truly becoming the darling of the press and of Wall Street.

BILL RETIRED FROM COMPAQ in 1987. He was the first founder to leave the company, and I was sad to see him go. Jim and I had been right when we decided in 1982 that Bill's contribution would be as important as our own. Bill's recommendation to sell only through IBM authorized dealers and his leadership of the sales and marketing teams was absolutely critical to our success. He was also a mentor to Swavely early on and later to Ross Cooley, an ex-IBMer who developed into a great vice president of sales after Bill left.

As 1987 drew to a close, we had little time to dwell on how amazing a year it had been. In achieving over $1.2 billion in sales, we set another record for the fastest any company had ever reached sales of a billion dollars. And we had done it in the same year our most dangerous competitor, IBM, had made its highly aggressive move in the PC market. Most of the world still did not understand how we did it, but they couldn't deny we were thriving in spite of IBM's repeated attacks that were expected to stop us.

The Compaq Portable 386—"pound for pound the most powerful computer on the planet."

In spite of our excellent financial performance, though, I continued to feel like there was a dark cloud hanging over us, and its name was Micro Channel. We met often to discuss the strategic issues it created, always using "The Process" to try to come up with a solution. But this problem was the toughest we had ever faced, and we really weren't making much progress. I was certain that if we were going to solve it, we were going to have to ask ourselves the right question.

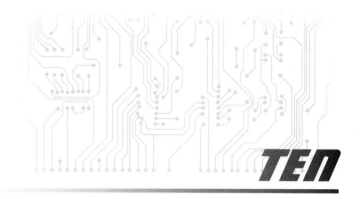

Compaq's Most Unexpected Decision

SHORTLY AFTER WE TOOK CARE of the immediate priorities that resulted from the PS/2 announcement, we began to look at our long-term alternatives. There were two obvious choices.

First, we could stay with the industry standard, continue to deliver performance advantages over the PS/2, and hope for a split market with two standards. When the PS/2 was announced, we had immediately and strongly stated our continued support for the existing standard, arguing that there was no need for 32-bit slots in the near term. This was the path most consistent with our public position. And our sales had continued to surge, despite IBM proclaiming its Micro Channel as the way of the future. But since technology was advancing so rapidly, we didn't know how long it would be before there was a real need for a 32-bit slot.

Second, we could reverse engineer the Micro Channel and gradually switch to delivering products compatible with the PS/2. I had consistently taken the public position that we would deliver Micro Channel products if and when the market demanded them. We had carefully kept this door open, and most industry observers believed we would eventually take this path. We knew it would be a lot more difficult this time, and we would still probably have to buy a license from IBM and pay it royalties. Most of our competitors were doing that. But to us, taking this route looked like a miserable existence, with IBM calling all the shots and allowing us no real future—certainly not as a technology leader. Even so, we kept this option viable.

We had expected IBM to introduce a new 32-bit slot but keep the 8- and 16-bit slots compatible with the existing standard. When IBM not only introduced a new 32-bit slot, but also eliminated the compatible 8- and 16-bit slots, we were amazed. Even though IBM said publicly this was necessary to achieve additional performance, we couldn't find any reasonable explanation, other than to reduce or eliminate competition from the clones. In doing so, IBM exposed an unexpected weakness we exploited in ongoing publicity battles. More important, IBM had opened up the possibility of Compaq doing "what IBM should have done."

A number of our competitors were suggesting that the industry develop a specification for an advanced 32-bit bus. Such an approach would offer the clear advantage of not having to pay IBM royalties for the use of its Micro Channel. Zenith had even proposed a formal Institute of Electrical and Electronics Engineers (IEEE) committee to develop a specification, and some work had been done on it. But we strongly believed such an approach wasn't viable. It would take far too long to get all the companies to agree, and almost certainly end in a compromise that would not be capable of winning the inevitable performance comparisons with Micro Channel.

As we looked down the road to a time when we would need a 32-bit slot on our PCs, the attractive choice was to create our own design. We believed we were the only company that knew how to achieve the advanced 32-bit performance needed to beat Micro Channel and at the same time maintain complete backward compatibility with the existing standard.

The situation wasn't quite that simple, however. Timing was a critical issue. If we waited too long before introducing our 32-bit slot, there was a great risk the Micro Channel would become so entrenched it couldn't be displaced. If that happened, we would be forced to adopt IBM's 32-bit bus late in the game, placing us at a significant disadvantage to our competitors who had made the move earlier. Whatever direction we decided to go, we needed to do it sooner rather than later.

Then there was the question, "Whose technology is best?" IBM had introduced some powerful new capabilities in its Micro Channel, at least on paper, so our 32-bit bus would have to be carefully designed to beat it.

All issues, however, paled in comparison to the really big one: How could we possibly come out on top going up against IBM? Sure, we had built a great reputation and were widely viewed as a technology leader. But Compaq was no IBM—not even close. Even if our 32-bit bus was better and faster, and even if the companion 8- and 16-bit slots were backward compatible, we had almost no chance to beat IBM in a head-to-head battle of this significance. Every time we even thought about going up against them, we could already hear the analysts saying, "Compaq's bus makes more sense, but I'll bet on IBM."

At this point, in the early fall of 1987, we hit a low point in our confidence. Our excitement over the likelihood that our annual sales would exceed $1 billion was severely tempered by a growing realization that we would probably have to switch to Micro Channel–based products. That prospect was a real downer.

Then one day in a product strategy meeting, someone asked the right question.

Our product strategy team gathers in the small conference room next to my office on the eighth floor of CCA-4, one of the new buildings on our headquarters campus. Present are Mike Swavely, Gary Stimac, Hugh Barnes, Steve Flannigan, Jim Harris, and myself, along with several others.

I open the meeting. "Let's get started. I'm having a real problem accepting the idea that we can't come up with a way to avoid having to switch to Micro Channel–based products. But I also know that we can't risk a direct confrontation with IBM over who has the best bus. So we need to get our heads clear and think out of the box."

I pause for a moment. With frustration sounding in my voice, I ask, "Is there any possible way the Micro Channel can be beat?"

Almost without thinking Swavely blurts out, "Sure, if the whole industry supported a better alternative. But they would have to do it pretty soon. Time is running out." He stops and thinks about what he just said, then adds, "I don't see any other possibility. The question is how to get the industry to move fast enough."

"OK," I say. "Let's focus on that for a minute. Think. Is there any possible way to get the industry to agree on a new bus quickly? What could we do to speed up the process?"

No one speaks for a while. Finally Barnes says, "One way to avoid the slowness of a committee process would be to design the bus ourselves. We're the only company with the technology to make sure it's completely backward compatible. And I think Paul Culley may be the best hardware architect in the industry, so he would be the right guy to do the design. Especially the new 32-bit part. If we could show our competitors a working bus with the right capabilities, it'd remove a lot of the risk."

Then Jim says, "One thing we've got going for us is that none of the others want to follow the Micro Channel either. But so far no one has come up with a viable alternative. If we could convince them our bus design would work, I think they'd all want to join in."

Swavely jumps in. "I'm sorry to play the devil's advocate, but won't everyone be afraid of becoming dependent on Compaq just like they are on IBM? We're their second largest competitor."

Stimac, starting to see some potential, responds, "Absolutely. So we'd have to set it up so they don't become dependent on us. We'd have to give up control of the design to the whole group, or to someone independent, like Intel. After the initial design is complete, of course."

Flannigan adds, "I think Microsoft would really get behind this. Bill [Gates] is very concerned about IBM gaining so much control. Really, all the software companies feel that way, so we could probably line up their support too."

Barnes is thinking about the design. "To have the same kind of capabilities as the Micro Channel will require several ASICs. Do we let everyone develop their own chips, or do we give those to them as well?"

I reply, "It might work either way, but it sure is a stronger offer if we take it all the way through chip development and let everyone buy chips from an independent supplier. And it speeds up the whole process. It should make their decision easier if we remove the cost and timing risk of chip development."

Everyone thinks about it for a moment. Suddenly my face lights up. "What if we do this behind the scenes?" I ask. "Imagine a press conference with all the leading PC companies except IBM standing together and announcing support for a 32-bit bus that is what Micro Channel should have been. That would be powerful."

Swavely says excitedly, "Now we've got something that could really work. That would be amazing." Swavely didn't get excited very often.

The brainstorming continues to gain momentum and goes on for another hour. Everyone is getting more and more pumped up as one objection after another is addressed. No one had entered the meeting with any inkling—even any hope—that something like this could emerge. But there it is, another completely unexpected decision that is solidifying in the moment. "The Process"—the way we turn critical decisions over and over until we find a clear path—is truly one of our secret weapons.

I summarize what we've come up with. "Compaq will create the specification for an advanced 32-bit bus and slot. Then our engineers will design the bus controller circuits and develop the ASIC chips that do the complex processing while maintaining compatibility with the existing 8- and 16-bit buses. We'll get Microsoft's support for our plan and to help us sell our competitors on the idea. They'll also develop the operating system software needed to handle the advanced bus. We'll allow a third party, probably Intel, to build the ASIC chips and sell them to our competitors on the same terms and availability that we receive. We'll not get any compensation or royalties on the chips that Intel sells.

"We'll visit the leading application software and add-in board companies that'll benefit from the capabilities of the advanced bus first and bring them in to support the plan too. I'm thinking of networking hardware and software, work group applications, file server applications, and advanced workstation displays and software.

"And we'll do this while keeping it a secret for as long as possible. Then, when it's all ready, we'll organize a major industry announcement where we'll introduce the new bus and the industry coalition that supports it. We'll all be viewed as equals."

As the team listens to the summary, the magnitude and significance of what we've come up with begins to sink in. We're planning to do what no other industry leader has ever done before: spend tens of millions of our own money to develop what could become the most advanced and valuable technology in the industry—and then give it away. Yet there's no doubt in my mind this is the only path that makes sense for Compaq. It's the only one that'll enable us to continue to be an industry leader far into the future—that is, if it works.

All revolutions start with broad discontent, but it always takes a spark to set them off. This meeting was the spark that started the revolt to save the PC open industry standard from becoming closed and controlled by IBM. The oddsmakers might not have viewed our chances very favorably at this point, but all the ingredients for success were there.

But first, we needed to be sure we could design a bus that would deliver what we were trying to achieve. The development project would require the efforts of many of our best engineers and programmers, but the first step was to design the basic architecture. We asked Culley to design an advanced 32-bit bus with features comparable to, or better than, those of the Micro Channel that would also be completely backward compatible with the industry standard. I wasn't surprised to find out he had already been thinking about the idea and was eager to move forward. This would be the most complex design he had ever tackled, and it was the most important.

Several weeks went by without a word from Culley. Then one day he walked into Barnes' office and said he had come up with a way to do it. After briefly discussing the design, Barnes leapt into action and put together a team to build a "proof of concept" to test the idea.

It was late December before the engineering team reached a point where they were sure. Barnes and Jim met with them to test their conclusions, and in the end they all agreed. Compaq could do it.

Next, we faced the question of whether we should continue to reverse engineer the Micro Channel as a fallback position. Should we continue until we were completely sure we could get enough of our competitors to sign up to support our advanced bus? There was a sense we were running out of time. It had already been nine months since the introduction of the PS/2, and would take at least six more to get the design ready and everything lined up for a major announcement. To move as quickly as possible, we had to shift a significant number of engineers off the Micro Channel project. It was time to put up or shut up.

JANUARY 5, 1988, 10:00 A.M.

Almost 300 people gather in a large meeting room at Compaq's headquarters in northwest Houston. Hugh Barnes, head of Compaq's desktop product development, has called together all the development personnel for an important announcement.

Barnes opens the meeting. "Thanks for coming. As all of you know, since last May we've been splitting our resources between new ISA product development and Micro Channel product development. The effect of this is that we're not able to move forward in either area as fast as we need to. We've decided that we can no longer afford to keep the resources split, so we have to pick one or the other."

Barnes pauses. "I want to make it clear that everyone in this room is critical to future product development. No one will lose their job."

He pauses again, and the room is so quiet you can hear a pin drop. The Micro Channel development team has made great progress toward creating a product that Compaq can take to market, but everyone knows if we end up following IBM down that path, we'll no longer remain the technology leader we've become.

When he speaks again, everyone is totally focused on his words. "After carefully considering the situation, Compaq's management has decided to shut down the Micro Channel development and focus all our resources on advancing the performance of our industry-standard products, including developing an advanced 32-bit bus that's compatible with the industry standard."

This time, there's a loud—almost deafening—cheer from the audience. This is the path they were hoping for. This is the path their hearts are really in.

When I heard from Barnes about the reaction of our development teams, I was really sorry I hadn't been there to share in their jubilation. We all knew there would be unexpected problems, but now we were fully committed to do whatever it took to make an advanced 32-bit bus succeed.

As soon as the proof of concept was complete, the rest of the project team immediately went into action. Swavely made a list of the PC companies to recruit for the alliance. We decided we needed two tiers in the alliance, because we wanted a manageable number of companies working together to make decisions. At the top of the list was Hewlett-Packard (HP). The list soon grew to over twenty companies.

Then we set up a meeting with Bill Gates at Microsoft to bring him in on the idea and to get his support. We had some concern that Gates might worry about damaging his relationship with IBM, but when Stimac, Swavely, and I visited Microsoft and laid out the plan, Gates didn't hesitate. He wanted this to work and would give it his full support, he said, including updates to Microsoft's operating systems. He would figure out how to deal with IBM later. When Stimac told him that we wanted to recruit a number of PC companies to join us, Gates said he had relationships with all of them and would get us in touch with the right person at each company. He would do everything he could to help us sell the idea.

Next, we set up a meeting with Intel to get its support. We were moving so quickly Barnes spent most of the night before the presentation in his hotel room, creating hand-drawn slides. As usual, Intel's executives were a little skeptical at first, but it quickly became clear to them that Compaq was going to make this happen.

We offered them the opportunity to be the supplier of ASIC chips for the advanced bus. They were interested, but a lot of contractual details had to be worked out before they were fully on board. They also had a relationship with IBM they needed to be careful not to damage. Before long, Intel was fully engaged and supportive. The company had become very close to us during the final stages of the 386's development and had seen us orchestrate the announcement of the first 386 PC. Its executives knew that if anyone could pull this off, it was Compaq.

We believed that our PC competitors would be the hardest to sell on the idea. We expected this group would be the most skeptical of both our probability for success and our real motive behind the plan. So we got everything lined up as perfectly as possible before we began to call them. The list of things to prepare was very long.

We needed to show very clearly how the advanced bus worked and how it compared with Micro Channel. It was also important to show the marketing aspects of the plan, which was a big part of convincing the competing companies the plan would work.

Our presentations had to be extremely professional and thorough; once we started talking to our competitors, we wanted to sign them up quickly. We also wanted enough of them signed to reach critical mass before IBM had a chance to hear about what was happening and mount some kind of interference. This might have been overkill on our part, but it was indicative of how important the success of this project was.

The first company Stimac and Swavely visited was HP. HP was very important because of its market position and technical capabilities. Even though we were prepared to do all the work ourselves, we could use some help from its resources. We also believed HP joining in would help sell the idea to the other companies.

In April 1988, Bill Gates, as he had offered, called Robert Puette, general manager of HP's PC division, and strongly suggested he listen carefully to what Compaq had to say. The meeting was set for early May, when the right people from HP could be there.

At the meeting, Stimac started the presentation with technical aspects of our proposal. Then Swavely followed with how he viewed the positioning of the advanced bus, the coalition, and the effect it would have. Many questions were asked throughout the presentation, but Stimac and Swavely were well prepared with solid answers. At the end of about two hours, the HP team was very interested and very impressed. But this was a big decision for them, as it had been for us. They needed time to consider it all.

When Stimac and Swavely got back to Houston, they met with the strategy team to discuss the results of the HP meeting and consider the timing of our next move. We concluded the odds were good HP would join us, so we decided to wait awhile to give its executives a chance to make their decision before meeting with the next company.

Over the next three weeks, there were many phone calls back and forth between Compaq and HP engineers, working out details and building a bond between the two teams. Finally, a call was set up between Robert Puette and me. Puette told me that HP was in, but he wanted to hear me commit that Compaq was going to make this a true industry coalition and did not plan on controlling the expanded bus. He had no way of knowing how strongly I was committed to doing just that. I replied, "Absolutely!"

In late June, Stimac and Swavely began to schedule meetings with other computer companies as fast as possible. With Compaq, HP,

Microsoft, and Intel committed to the coalition, we were now confident the others would be relatively easy to convince that the "bus" was about to leave the station and they needed to be on it.

As it turned out, the other companies weren't quite that easy to convince. Although the presentation was compelling, hearing the idea for the first time was somewhat shocking. All the companies needed time to absorb what they heard and then talk to their decision makers. Some of them had already made strong commitments to support the Micro Channel and had to think through how this new coalition would affect those commitments.

Dell Computer, for example, was touting the fact that it had hired the IBM engineer who had led the design of Micro Channel to lead its own efforts. Despite our assurances, some companies were concerned that Compaq would end up dominating this area, similar to the way IBM was dominating the Micro Channel. Many of the companies required more conversations to clarify various points.

Decisions to join the coalition trickled in slowly through July and early August. We had set the announcement date for September 13, 1988, and I began to worry we wouldn't have enough support to be credible. We decided on a cut-off date for companies to be included in the press conference, as a way to nudge those hanging back to go ahead and join us. We ended up with nine companies large enough to be part of the founding group. In addition to Compaq and HP, they were: AST Research; Epson America; NEC; Olivetti; Tandy; Wyse; and Zenith. We would become known as the "Gang of Nine."

One of the many issues that had to be addressed was the name of our bus. The name needed to convey advanced capabilities but, even more important, compatibility with the existing industry standard. The name we decided on was "Extended Industry Standard Architecture," or EISA. The acronym was short and easy to pronounce, and the four words communicated exactly what EISA was all about.

Everything was falling into place as the day of the announcement drew near. When word of the press conference began to spread, many of the computer companies that had been on the fence, including AT&T and Dell, agreed to join us. They were too late to be included in the press conference panel, but they wanted to be counted in the alliance. A week before the announcement, there were more than 60 companies supporting the alliance; by the day of the announcement, the number had grown to 80. Every company in the industry except IBM wanted this coalition to succeed, and a strong feeling had developed that it would.

SEPTEMBER 9, 1988, 6:00 P.M.

It's the Friday before the press conference and I'm on the West Coast finishing meetings scheduled around a convention I am attending. I'm in a discussion at a San Francisco hotel going over some of the final details for the press conference when I'm handed a note. Bill Lowe, president of IBM's Entry Systems Division, is on the phone. I had placed a call to Lowe's office a few days ago, but didn't expect him to return the call. Without mentioning the caller's name, I look around the room and say, "I need to take this call." I ask that the call be put through to a phone in the hallway outside the meeting room.

When the phone rings, I pick it up. "Hello."

"Hello, Rod. This is Bill Lowe."

"Hi, Bill. Thanks for returning my call. I'm sure you've heard that there's an announcement next week." I pause briefly, trying to decide how much to tell him. "A number of PC companies are joining together to support an advanced 32-bit bus that's compatible with the existing standard. It's open to all companies that want to join in, so I wanted you to have the opportunity to do so."

Lowe replies, "I understand. IBM isn't interested in joining, but I want you to know that we'd welcome you as a licensee of the Micro Channel. We'd be willing to work with you to help you get to a Micro Channel product quickly."

For a moment, I don't reply. I'm dumbfounded. I'd never expected Lowe to turn the offer around. Nine months ago we couldn't have passed this up, but now we're on a much better road. I almost say, "Your timing is a little off." But instead I tell him, "Well, thanks for the offer. We'll think about it."

Lowe responds, "OK. Let me know what you decide. And with regard to your announcement, I think you're making a big mistake."

I resist the temptation to argue the point.

After I hang up, I stand there for a moment, wondering if Lowe is serious or just taunting me. At this point, I wouldn't have changed to the Micro Channel if Lowe offered the license free with no royalties. EISA is the right thing for our customers, and it will significantly outperform Micro Channel.

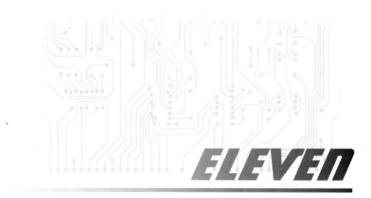

Compaq Leads the Revolt

Several hundred people have squeezed into a large meeting room at the Marriott Marquis Hotel in New York City for a press conference. The event has created tremendous interest and excitement throughout the PC industry. This morning's financial news television shows and newspapers reported on what is about to take place.

Sitting at a table across the front of the room are the CEOs or general managers of nine of the leading PC manufacturers. Richard Shaffer, a highly respected PC industry consultant and former *Wall Street Journal* technology editor, steps to the podium. He welcomes everyone, then introduces the nine members of the panel in alphabetical order by company name: Safi Qureshey, AST Research; Rod Canion, Compaq; Eugene Kunde, Epson America; Robert Puette, Hewlett-Packard; Richard Underwood, NEC Information Systems; Franco Agostinucci, Olivetti; John Patterson,

Tandy; Phillip White, Wyse Technology; and John Frank, Zenith Data Systems.

Next, I'm invited to the podium to make the announcement. "I'd like to begin by explaining the common objectives of these major manufacturers of industry-standard personal computers as they relate to EISA, that is E I S A, the acronym for Extended Industry Standard Architecture. First, this group is committed to opening the way for continuing growth of industry-standard personal computers. At the same time, we're committed to maintaining compatibility with the existing standard for systems and peripherals. With this EISA standard, we will work to maintain open, industry-standard PC platforms, giving third parties the confidence to continue adding value with their innovative new products. Finally, this group will continue to elicit the support of the other manufacturers in the PC industry, to strengthen the EISA standard and continue expanding the benefits it provides.

"To understand the significance of these commitments and the importance of today's announcement as a major evolutionary step in the maturing of the PC industry, let's look back at the progress of the industry standard since its inception."

I explain that the industry standard started with the IBM PC and then continued to evolve as new technologies were added while maintaining compatibility with the existing base.

"This smooth progression of advances resulted in the maintenance and expansion of a unified industry standard accommodating all the technologies within that standard, from the earliest to the most recent.

"With each new technology, there's always been compatibility with the previous generation. That meant users were always able to use the original applications and run the original peripherals they had purchased as they made a smooth transition to new, more powerful PCs.

"This enduring and expanding industry standard has resulted in the availability of thousands of application software packages, hundreds of 8- and 16-bit expansion boards, and hundreds of peripherals. Just as important, literally dozens of manufacturers of industry-standard PCs are constantly competing to deliver new technologies to users faster, more efficiently, and more cost effectively.

"This industry standard has resulted in a phenomenal explosion in the acceptance and use of personal computers—what we now know as the personal computer revolution. We've seen users flock to industry-standard architecture PCs literally by the tens of millions."

Next, I point out that PCs based on industry-standard architecture far and away dominate the market, currently accounting for 66 percent of all PCs sold to businesses through computer dealers. Micro Channel makes up only 20 percent, and Apple about 13 percent.

"The industry-standard architecture will continue to strengthen and evolve. In the future, we'll see faster 386 microprocessors and eventually a compatible 486 microprocessor. We'll also see new, compatible versions of the standard operating systems, including DOS 4.0 and OS/2 1.1.

"You can see that the future will bring higher-speed peripherals that'll require a new 32-bit bus that is compatible with the 8- and 16-bit buses that came before. Today, together, we're introducing this new 32-bit Extended Industry Standard Architecture bus.

"Today's application software and operating systems do not use all the bandwidth available on the existing ISA bus. This applies to OS/2 and coming applications on OS/2 as well. While both the Extended Industry Standard Architecture and Micro Channel Architecture offer the potential for increased performance in the future, they simply do not impact the performance of today's applications.

"But in the future, after we see development of peripherals and software that can take advantage of these advanced bus architectures, application performance will improve. EISA will be capable of delivering performance significantly higher than both today's ISA and Micro Channel."

Then I list the features included in the EISA bus specification and compare the performance of EISA with IBM's Micro Channel and Apple's NuBus. From a specification standpoint, I make it clear that EISA will significantly outperform both other alternatives, and is the only one compatible with all existing industry-standard plug-in boards and peripherals.

"In summary, today's introduction of the Extended Industry Standard Architecture for personal computers is an important milestone in the evolution of our industry. EISA sets the standard for a new, advanced 32-bit bus for PCs. Most important, since it is completely compatible with the existing industry standard, it protects the investments of today's millions of PC users worldwide. The EISA design is open to all developers, making it a safe path to follow when designing future products. And we'll see the first of these products incorporating EISA begin to emerge in late 1989.

"EISA is a true industry standard—just like its predecessor. There is broad support from PC manufacturers who will incorporate the EISA bus into their new industry-standard personal computers. There is broad support from system software vendors, who are already working to ensure that new operating systems will take advantage of the higher performance of the 32-bit bus while maintaining compatibility with today's applications. And numerous third-party peripheral and board manufacturers already are developing new products that'll work with the new bus. Availability of logic chips required to support the new bus helps ensure support from the broadest range of suppliers. And because EISA is compatible with their installed base of PCs, and is painless to adopt, we are confident that EISA will achieve broad support from the user community as well."

When I finish, I receive a moderate amount of applause. The audience is weighing the argument for the existence of a new bus, and especially its probability of success. The "Gang of Nine" anticipated this response and included other industry leaders in the announcement.

Next to the podium is Albert Yu, vice president of Intel, who proceeds to express Intel's strong commitment to EISA. "This event exemplifies the industry's determination to create and extend standards."

He says Intel will produce the chips that both PC and peripheral manufacturers will use to simplify the design and lower the cost of products that use the EISA bus.

Up next is Steve Ballmer, vice president of System Software for Microsoft, who expresses Microsoft's strong support of EISA. "It's important that we provide system software products that fully support EISA-based machines. In the future, our systems software products will evolve to support new, high-speed, 32-bit devices, such as high-performance disk subsystems and network inter-faces. So the availability of the EISA 32-bit slot capability is some-thing that we intend to fully exploit.

"The potential for EISA-based machines to employ multiple processors is also important. This capability will also be exploited by our systems software. The most likely first use of such a capabil-ity will be by the OS/2 LAN Manager, which will seek to exploit the performance advantage of a multi-processor system in a local area network server configuration." He explains that the company will not only provide a version of MS-DOS and OS/2, but also Xenix, Microsoft's version of Unix.

Shaffer returns to the podium and announces that in addition to leaders from Intel and Microsoft, also present are CEOs from

Digital Communications Associates, Novell, and 3Com—three companies that provide communications and networking hardware and software. He explains that they'll be up shortly to answer questions from the audience.

He opens the floor to take questions from the audience for the "Gang of Nine." Many questions are aimed at trying to determine if nine different competitors are going to be able to work together long enough to get the new bus accepted. The "Gang" is convincing in its resolve to make EISA a true industry standard.

Then Shaffer announces that the industry-support panel will replace the PC-manufacturer panel at the table. He reintroduces Yu of Intel and Ballmer of Microsoft, along with James Ottinger of DCA, William Krause of 3Com, and Raymond Noorda of Novell. More questions from the audience are taken until there is a discernible pause. Finally, Shaffer says all the members of both panels are available for interviews and closes the formal part of the event.

The press conference went very smoothly. Although the Compaq team had done almost all the work to this point, we were very careful to make it clear that this was a coalition of equals, with every company heavily involved. In reality, we had contributed about 90 percent of the effort, HP about 9 percent, and all the others together, about 1 percent.

It was disappointing that neither Bill Gates from Microsoft nor Andy Grove from Intel attended the event. Even though the chairmen of both companies had given speeches at Compaq's 386 launch in September 1986, apparently both felt this would have been a bigger problem for their relationships with IBM. Both had made it clear to me they wanted the coalition to succeed, and sending other top executives showed strong support.

On September 14, the day after the press conference, the coalition ran a jointly sponsored full-page ad in the *Wall Street Journal*. In giant bold type running down the middle of the ad was the punch line: "OVER THE PAST 8 YEARS, BUSINESS HAS INVESTED MORE THAN $100,000,000,000 IN INDUSTRY-STANDARD PERSONAL COMPUTING. TODAY IT JUST PAID OFF." We had been beating up on IBM about abandoning all the industry-standard hardware and software investment ever since the PS/2 announcement. But this ad was like shouting directly in its face.

Remarkably, IBM decided to make an announcement on the same day—and at the same time—as the EISA press conference. IBM's announcement was held at its headquarters in midtown Manhattan. One part of its presentation focused on the PS/2 Model 30 286, a low-end computer that used the industry-standard bus instead of its own Micro Channel. As a low-cost, low-end product, the computer made perfect sense. But announcing it on the same day as the EISA announcement didn't. It gave the press and analysts a tempting comparison that was unfavorable to IBM.

IBM had almost always followed a policy of not commenting on competitive announcements, but that day, its executives spent much of the Q&A period defending Micro Channel. Although they knew very few facts about EISA, that didn't stop them from attacking it. The IBM spin machine was running at full speed.

Since there was no World Wide Web at the time, the first media outlet to carry the story was television, in particular, business news programs such as *Moneyline* on CNN. In one of the first segments to appear, Lou Dobbs opened with: "Big Blue is doing some embarrassing backpedaling in marketing. The backpedaling brought on as many of its competitors mounted an historic challenge to IBM's role as creator of standards for the PC industry." Dobbs then introduced Steve Young to report the story.

The video cuts to a production line of soft drink bottles moving swiftly along. Young says, "What, you ask, does New Coke have to do with IBM?" He explains that when IBM introduced a new personal computer "flavor," the PS/2, it rendered useless $12 billion worth of software that businesses had purchased. That caused IBM's market share to drop significantly, he noted, and as a result other computer companies were selling two-thirds of the PCs bought from dealers. "Faced with overwhelming resistance to the flavor of 'New Compute,'" Young says, "IBM today quietly, very quietly reintroduced 'Classic Compute.' It looks like the new computers, but has a souped-up version of the old AT brain. Is it an effort, as many analysts contend, to regain market share by bringing back 'Classic Compute'?"

Later Young suggests that IBM may continue to lose market share because its competitors agreed to use the newly announced EISA bus instead of IBM's. It cuts to a shot of me speaking at the EISA press conference saying that because EISA is compatible with the existing industry standard, it protects the investment of millions of PC users worldwide. Young wraps up the report with, "Some analysts think it was a day that left Big Blue looking black and blue."

I couldn't believe it when I saw this news report. I thought the story couldn't have been more positive toward us if I had written it myself. I had used that "New Coke" analogy during the April 1987 interviews just after the PS/2 was announced. It had worked well before and apparently was an irresistible analogy for the media again.

Other news reports were similar, but not quite that positive. They included Neil Cavuto on the *Nightly Business Report*, Doug Ramsey on the Financial News Network's *Business This Morning*, and Bob Jamieson on NBC's *Before Hours*, among others. The common theme: The industry IBM had led for seven years was in revolt, and it didn't look good for IBM. They all pointed out that the company had been losing market share since the introduction of its PS/2 and reported our EISA announcement could accelerate the trend. Most of the reports quoted IBM executives. While they each had smooth deliveries, they all came across as defensive.

Soon there were major articles in all the business newspapers and magazines. For the first time in Compaq's history, it seemed to me the analysts' quotes were generally negative toward IBM and favorable toward Compaq's, and EISA's, chances for success.

The longest article was in the *Wall Street Journal.* On September 14 in an article titled "Nine Firms That Make Personal Computers Gang Up Against IBM," Paul Carroll and Michael Miller wrote, "Nine big personal computer makers are doing what computer companies have dreamed of doing for decades: ganging up on International Business Machines Corp. They are joining forces in an audacious attempt to wrest away from IBM the power of setting the standard for how personal computers are designed, and they seem to have a chance of succeeding."

Several analysts and consultants were quoted in the article. Aaron Goldberg, vice president of IDC, noted that this was the first time the compatible industry was deciding its own future. Tim Bajarin, a consultant at Creative Strategies Research International, said that if there was ever a chance to force IBM to lose its dominant industry position, this was it. Jonathan Yannis, a Gartner Group analyst, noted that there was a "better-than-even chance" for the coalition to consign the Micro Channel to the subset of buyers who go with IBM simply because it's IBM. And Charles Wolf, an analyst at First Boston Corp. who had been pessimistic about Compaq after IBM introduced the PS/2, did an about-face and predicted Compaq would be the big winner if the effort was successful.

The article concluded that Compaq had the bulk of the risk: "Because in recent years it has been fast building a reputation for the best technology in the industry, it has the most to lose. And few companies that are No. 3 in an industry—Compaq's position behind IBM and Apple—have achieved the sort of influence that Compaq is seeking."

Its comment on the coalition's impact on IBM was chilling: "For IBM, the gang's announcement yesterday is at best a dust storm of confusion, and, at worst, a dagger to the heart of its PC strategy."

Similar, but shorter, articles appeared in the *New York Times*, *BusinessWeek*, the *Financial Times*, and *USA Today*. In a September 14 *USA Today* article titled "IBM rivals challenge its standard," Katherine Hafner closed with, "A group of manufacturers that prospered by aping IBM's standards may succeed in calling the tune on personal computer design. And Big Blue could end up cloning the cloners."

Peter Lewis, in a September 11 *New York Times* article titled "A Big Battle Over Computer Buses," wrote, "Just to confuse things further, IBM will also make an announcement Tuesday, introducing, among other products, the latest model of the Personal System/2 family—a Model 30 that uses the old AT chip, the 80286, and—surprise!—the old 16-bit AT bus architecture. That makes it sound as though IBM is making a U-turn into oncoming traffic."

Coincidentally, the Salomon Brothers Microcomputer Conference was held in New York the day after the EISA announcement. With both IBM's Lowe and me speaking, there was no way to avoid contact. Actually, I was glad to have the opportunity to set the record straight with regard to the spin and misstatements coming from IBM. If the press thought I had suddenly grown claws when I became more aggressive after the PS/2 announcement in April 1987, they were now going to see me crank it up another notch or two.

I was sitting in the audience when Lowe gave the luncheon address. He hadn't seen EISA's specification, he said, so he couldn't make an authoritative comparison between the two buses. But he added, "We think Micro Channel is more robust." He failed to say specifically on what he based that assertion. Later, he said, "We've protected our customers' investments in that Micro Channel ensures you can take advantage of new technologies for years to come."

When a reporter approached me after lunch, I was ready. IBM had been promising that machines with Micro Channel would be able to take advantage of "new technologies" and "new applications" for almost a year and a half since the introduction of the PS/2, and still had nothing to show. Also, when our engineers were analyzing Micro Channel and developing EISA, they discovered a severe weakness in the Micro Channel–based PS/2 machines IBM had already shipped. Those computers lacked a memory cache for the processor, which meant that most didn't have enough performance to deliver on IBM's promise.

I told the reporter, "I think his answers were really misrepresentations.... Without cache memory for the processor, those 1.5 million Micro Channel machines will be unable to take advantage of applications that require higher bus bandwidth." The cracks in IBM's story were starting to get bigger.

A week later, Dataquest Research International held a PC Conference in Napa Valley, California. The scheduled topic was "Visions of the '90s," but the major subject of interest centered on EISA.

An article in *Computer Reseller News* on September 26 titled "Dataquest Conference Focuses On EISA" pointed out that almost all the conference attendees supported EISA. Ed Anderson, chief operating officer of ComputerLand Corp., said "The new EISA standard corrects (an issue) that has left the customer stranded in terms of investments" and that EISA was what Micro Channel Architecture should have been. MicroAge Computer Stores Inc. chairman Alan Hald called EISA "a natural extension of Compaq's concept of continuity."

Over the next two weeks, there was a lot of coverage in PC-industry publications, including *Information Week*, *MIS Week*, *Computer Reseller News*, *Computer World*, *Computer & Software News*, *PC Week*, and *InfoWorld*. Overall, the coverage was evenly mixed, with some articles strongly supporting EISA while others strongly supported Micro Channel and cast doubt on EISA's survival.

A main argument against EISA was one IBM had started. By the time EISA machines actually arrived, IBM said, Micro Channel would already be entrenched. This theory overlooked the fact that almost none of the installed Micro Channel machines were suitable to run applications requiring an advanced bus, due to the lack of a memory cache for the processor.

Another argument came from an IBM engineer, who said that his team had looked at doing a bus like EISA, but rejected it because of inferior performance. In the past, IBM could get away with making a statement like that without being challenged. But we had built a strong reputation as a technology leader and could make credible counterarguments.

Interestingly, both *PC Week* and *InfoWorld* published editorials that cut to the bottom line. An editorial in the September 26 issue of *PC Week* titled "EISA: More than Just Another Bus" stated, "The new 32-bit bus design, EISA for short, deserves a thunderous round of applause...EISA is nothing less than a path not merely for survival, but prosperity. It's the sanest, strongest answer to IBM's pullout of the PC mainstream last year in favor of doing something proprietary (Micro Channel Architecture) and threatening anyone who tried to clone it." The copy of the editorial that I have didn't include the author's name and I wasn't able to find it.

William F. Zachmann, Vice President of Market Research for International Data Corp., wrote an editorial for the September 29 issue of InfoWorld titled "What's So Confusing? EISA Extends Capabilities of Industry-Standard AT Bus." In it, he addressed "the supposed confusion" introduced into the marketplace by EISA.

I nearly fell out of my chair when I got to the fourth paragraph. It said, "I'm not sure whether the folks writing this stuff simply are confused easily themselves, working for IBM's public relations department, or just downright stupid....However much the folks at IBM would like everyone to believe this sort of nonsense, the idea that EISA confuses matters is ridiculous. EISA is not a 'third' standard at all. It is simply a straightforward extension of the *de facto* industry-standard AT bus..."

Later, he goes on to say, "The supposed advantages of the MCA have been more a creation of the imagination of IBM's public relations and advertising departments than of IBM's engineering departments. The MCA has failed to demonstrate any real functional or performance advantages...From the start, the MCA was designed more with IBM's interests than customers interests in mind. It is small wonder then that many users have rallied behind the slogan: 'MCA—Just Say No!' EISA merely strengthens the case for those who prefer to continue enjoying the benefits of intense competition among vendors that an open architecture makes possible."

Although the battle was far from over, the momentum had definitely shifted in the direction of the open industry standard.

Zachmann got it exactly right: All of IBM's spin and "FUD" (fear, uncertainty, and doubt) couldn't confuse something as simple and straightforward as an extension to the existing industry standard. If there was any confusion created among customers, it was in those who were about to choose Micro Channel. Most were buying it because of the IBM brand. They may have paused to consider how much better it would be if they could continue buying machines compatible with their installed base. And anyone leaning toward industry-standard machines now had one more reason to stay with the standard.

Even with the first EISA-based machines a year away, we had effectively thrown a roadblock in front of IBM's march toward proprietary control of the industry. Although the battle was far from over, the momentum had definitely shifted in the direction of the open industry standard.

Looking back at all this, I'm still amazed we had the guts to attempt it, and by how well it actually worked. There was definitely a lot of risk involved, but from our perspective, long term it was the least risky path of all. Even though we were experiencing record sales growth during the second half of 1987, I was convinced that IBM would eventually win if customers weren't shown a better choice. And fortunately, our success hadn't deluded me into believing we could win a one-on-one battle against IBM. I didn't want an Alamo-like last stand.

But no one could have anticipated a solution like the one we came up with. Then, to minimize the overall risk, we identified each individual risk area and devised a plan to offset as much of the risk as possible. After that, it was simply a matter of excellent execution of each part of the plan. All the tens of millions of dollars and countless hours of work we invested in pulling this off were clearly worth it.

COMPAQ'S SALES CONTINUED TO BOOM throughout 1988, and we ended the year beating analysts' expectations by delivering over $2.1 billion in sales. We continued to introduce new products throughout the year, including three new desktop models and our first entry into the laptop market. All made significant contributions to our revenues. We also stayed ahead of IBM as the performance leader in the 386 segment by introducing the Deskpro 386/25 on June 20, 1988. Although IBM announced a competing 25-megahertz, 386 product at about the same time, we were shipping our product in volume immediately while IBM didn't begin shipping its products for about three months.

Plus, the Compaq Flex Architecture enabled us to significantly outperform IBM's 25-megahertz 386 product, even though its machines included the Micro Channel.

In the same June 1988 announcement, we also introduced a lower-cost 386 model, our Deskpro 386s, the first computer to use the Intel 80386sx chip. We positioned this product to address a broader segment, the upper end of the 286 market, by pricing it lower. Sales took off quickly.

Just a week after the EISA announcement, on September 19, 1988, we held another press conference where we introduced the Deskpro 386/20e, a smaller, lower-priced version of the original Deskpro 386/20 launched a year earlier. This introduction expanded our 386 product line to include five different performance ranges at five different price points. The 386 market had been ours ever since we were the first to introduce a 386-based personal computer. This broad product line enabled us to maintain our leadership position in that important market segment.

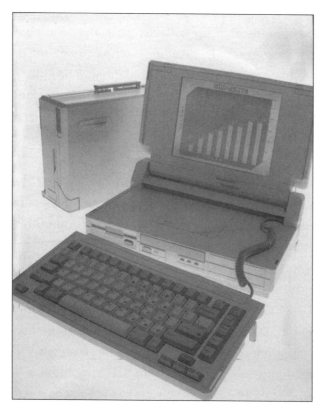

Compaq entered the laptop segment in October 1988 with the SLT/286.

Then a month later, on October 17, we entered the laptop segment with the SLT/286, a 14-pound, battery-operated computer. Compaq had been criticized by analysts for not going into laptops earlier, projecting we would not be a factor in that market because of our late entry. We had answered these criticisms by pointing out the technology at that time did not allow a laptop to be "full-function," a term we often used to describe our portable products. The SLT/286 was the first truly full-function laptop, delivering a high-performance 286 processor, high-resolution LCD display, and a hard drive. It even had a removable keyboard.

The product received great reviews and achieved the highest initial order rate of any product in Compaq's history. It took nearly six months to begin to catch up with demand. By the time we did, the SLT/286 had become the best-selling laptop through the dealer channel. And we had stayed true to our fundamental product philosophy by waiting for the right time to enter a new market. Counter to the analysts' predictions, we quickly became the leader in that market too.

IN JANUARY 1989 I became aware of a problem that was brewing with one of our largest authorized dealers, Businessland. It had always pushed us for bigger discounts than its competitors were getting, and it had been a struggle all along to keep a good relationship. We had been very careful to treat all our dealers fairly and consistently. But now I was told that Businessland was offering its sales personnel a $100 bonus over its normal commission for each IBM PS/2 they sold. We began getting reports from some loyal customers that when they went into a Businessland store and asked about a Compaq computer, the salesperson had disparaged Compaq and tried to sell them a PS/2. This was very troubling to me. I told Ross Cooley, our vice president of sales, to be very clear to Businessland's management that we could not tolerate it continuing that practice. I asked that I be updated regularly on where things stood.

In early February we had a meeting to specifically focus on what to do about Businessland. I found out that it had continued its bonus program, and the "bait-and-switch" reports were getting worse. I also found out that Dave Norman, Businessland's CEO, had recently told analysts that he planned to focus most of their technical support behind the IBM PS/2. We concluded that Businessland was putting extreme pressure on us to cave into its demands for special treatment, something we couldn't do without creating major problems throughout the rest of our authorized dealer network.

The issue was what to do about it. It was clear to me that we would be better off without Businessland under these circumstances. It would be painful to terminate our relationship with them, but we would be better off in the long run if they refused to stop what they were doing. We decided to identify and work through all the issues associated with the termination process and prepare a letter of termination before sitting down with its representatives for a showdown.

FEBRUARY 21, 1989, 11:00 A.M.

Mac McLoughlin, Compaq's Vice President of Sales for the Western Region, has picked up Cooley, Swavely, and me at the San Francisco airport and we are on our way to a meeting at Businessland headquarters. We discuss our plans for the meeting with McLoughlin, but the mood is somber. We are prepared to do something we really don't want to do.

We arrive at our destination and McLoughlin leads us into the lobby of the building, then remains there with Cooley while Swavely and I are escorted into a large conference room. We introduce ourselves to several Businessland executives and then engage in small talk while we wait about ten minutes for their CEO to join us.

Finally, a tall, thin Dave Norman strides into the room and we exchange formalities. After sitting down, we quickly get down to business. I say, "Dave, I want you to know that we appreciate the way you have worked with us over the years. I'm really disappointed that we're having difficulties now. What can we do to get our companies back to a better working relationship?"

Norman replies, "We have decided to focus our resources behind a few companies that are willing to partner with us and treat us like the unique company that we are. IBM has already done so, and we want you as well."

I respond, "You are very special to us, but you know that we have committed to all our dealers that we will not give anyone special discounts that would give them an unfair advantage."

He says, "We aren't like your other dealers. We give our customers better technical support and therefore deserve to be compensated for it."

The other participants join in and we all discuss issues related to their uniqueness and our commitment to our dealers. Finally, things quiet down. I look at Norman for a moment, trying to think of how I am going to handle this. Then I say, "Dave, the bottom line is that we are not going to give you deeper discounts. We have thought this through carefully and decided that we would be better off to not be in your stores than to continue with you incenting your sales personnel to switch customers from Compaq to IBM."

Norman stares at me and his face begins to turn red. "Do you think you can come in here and threaten me? Let me tell you something. Some IBM executives were in here a few weeks ago and told me they couldn't give us better prices. I threw them out. And two weeks later they came back with their tail between their legs begging us to take better prices."

With a very measured voice, I say, "Dave, I'm not threatening you. I'm simply trying to get you to understand this situation from our perspective."

"If I can make IBM jump like that, why should you be any different?"

Norman's face is getting redder. I look at him for a long moment, then turn to Swavely on my left. He hands me a letter from his briefcase. I turn back to Norman and say, "I'm sorry, Dave, but I see no alternative but to end our relationship. This is a letter giving you official notice that we are terminating Businessland as a Compaq Authorized Dealer."

I slide the letter across the table to Norman. He takes it and silently looks it over. Then his face turns so red I think he is going to explode. Without a word, he gets up and stomps out of the room. The Businessland executives are in shock, looking at each other but not saying anything. Finally, Swavely and I get up.

I say, "Thanks for your time," and we walk out.

McLoughlin drove Cooley, Swavely, and me back to the airport for our flight back to Houston. On the way we filled them in on the meeting. Cooley asked if there was a chance to reconcile, and I replied that there probably wasn't. We had decided before we came that if we actually gave them the letter, we couldn't go back. It would be very difficult to trust that they wouldn't be constantly looking for some way to get even.

When we got off the plane in Houston, there were several phone messages waiting for us. Several executives from Businessland had called several different Compaq executives, saying there was a misunderstanding and that we needed to work it out. But there was no turning back.

Compaq represented about 15 percent of Businessland's sales, while it accounted for about 7 percent of our sales. Both companies suffered only slightly short term. Our other dealers were so excited that we had stuck to our commitments to them that they became very loyal to Compaq. They quickly made up the difference and more.

It wasn't quite the same story for Businessland. In pushing IBM over Compaq it was effectively betting on PS/2 and Micro Channel. We continued to take market share away from IBM and, after our new products launched in late 1989, Businessland began to lose market share as well. Then the recession of 1991 caught it in a weakened state and it almost had to file for bankruptcy. Businessland was bought by JWP, Inc. in August 1991 for a small percentage of its previous value.

With the Businessland matter behind me, I remained focused on completing the task of defeating Micro Channel throughout 1989. I knew we weren't there yet. We still needed to deliver EISA-based products that clearly proved their compatibility and performance superiority over Micro Channel. Everything seemed to be progressing well until mid-March, when an unexpected problem popped up.

Intel Begins to Drift

Hugh Barnes, vice president of engineering, has requested an emergency meeting with our strategy team. As soon as everyone is seated, he starts. "I believe we have a serious problem with Intel. The 486 chip is behind schedule and falling further behind."

I ask, "What's the problem? Is the chip becoming too complex?"

"That's part of the problem. But the real issue is, they don't have their best people working on it. They've started a project to design a RISC (Reduced Instruction Set Computer) processor and assigned several of their most experienced chip designers to handle that instead of the 486. Apparently they've prioritized the RISC project ahead of the 486."

I shake my head slowly. "What are they thinking? How could getting into the tiny market for RISC chips be more important than protecting their monopoly with the industry standard?"

"I don't know. Maybe Scott McNealy's rhetoric is affecting them."

157

The CEO of Sun Microsystems is getting a lot of press with his pronouncements about how much faster his SPARC RISC processor is than Intel's CISC-based, X-86 processor (each Complex Instruction Set Computer instruction can execute several low-level operations). He's been saying that the next-generation SPARC will be two to three times faster than Intel's next-generation 486.

"Is there any truth to McNealy's claims?"

Barnes thinks for a moment. "There's some speed advantage to the RISC approach, but I doubt it's as much as he claims."

I look around the room. "It appears we're going to have to convince Andy Grove [president of Intel] that the 486 is more important than chasing the RISC market. Does anyone have an idea about how we can do that?"

Stimac, vice president of systems engineering, speaks up. "I think we can get Bill Gates to help. He has a vested interest in the industry standard staying on top."

I reply, "OK, contact Gates and see if he's willing to help. Hugh, see if we can get a meeting set up with Andy. I can fly out there, but I think we should meet somewhere off-site from his office. And let's do this as soon as we can."

I sit there and think for a minute. I'm wondering why Intel doesn't understand the power of the industry standard and the importance of keeping it at the forefront of performance. As if fighting off IBM wasn't enough, now we have to try to keep Intel from letting the 486 chip slip further behind. We can't afford to lose the incredible advantage the industry standard is giving us.

I realized the meeting with Grove would be pivotal for the future of Compaq and the open industry standard. I had to find a way to get him to fully understand and appreciate the open industry standard's strengths and benefits. Then it hit me: While I had talked about the importance of the industry standard for many years, I had never fully articulated all its benefits and how they worked together to deliver such a significant impact. I had to get the benefits straight in my own mind before trying to convince Grove.

I spent some time writing down my thoughts, and then arranged them in order of importance. I looked at my completed list and thought it was compelling:

1. guaranteed access to the latest software and peripherals
2. rapid introduction of key innovations and advanced technologies—higher performance, higher capacity
3. lower prices—continual downward pressure due to intense competition
4. broadest range of brands and product choices ever
5. ability to use existing software and plug-in boards when upgrading, even to a different brand PC

The combination of all these working together was stronger than any one individual benefit, and gradually created a chain reaction in the PC market. As prices went down and performance went up, more and more users were drawn in. Once they were hooked on the increased productivity they achieved, they were often attracted to upgrade to new computers with more performance and innovative new features, especially since upgrading was such an easy process.

As the rate of market growth increased, Intel and Microsoft began to generate incredible profits, which they wisely plowed back into rapidly developing advanced technologies. As these came to market, more users were drawn in, and then the cycle repeated.

These benefits created such a strong attraction to industry-standard PCs that even IBM couldn't easily pull customers away. However, if Intel couldn't deliver a high-performance 486 processor on time, it would give IBM, Sun Microsystems, and others the opportunity to tear down the open-industry-standard empire.

I knock on the door of a small hotel suite in Silicon Valley. Bill Gates opens the door and invites me in. Already in the room are Andy Grove and Gordon Moore, president and chairman of Intel respectively. The four of us exchange greetings and then sit down around a table.

I jokingly ask, "Who called this meeting?"

Grove looks at me seriously and says, "I think you did."

"Well, I guess I did. Thanks for coming. I'll get straight to the point." I look directly into Grove's eyes and say, "My team and I believe you are giving a higher priority to developing a RISC chip than to completing the 486, and the result is that the 486 is slipping further and further behind."

Grove is ready with a response. "Both chips are very important to Intel. We have excellent teams working on both. The 486 has had a few problems, but it's back on track."

I reply, "Look, Andy. I'm not going to try to tell you how to run your business. I know getting into the RISC chip business is important to you. But as you know, Compaq's engineers have worked closely with yours for a long time now, and they're telling me your very best people are working on the RISC chip. What I want to do is make sure you understand how important the 486 is to all our futures."

Grove comes right back at me. "The microprocessor market is much bigger than just the X-86 segment. We have to be able to develop more than one new product at a time. The problem is McNealy's hype seems to be working, because we see a lot of computer companies evaluating the Sparc chip. If we don't provide those companies with a viable alternative soon, Sun will end up dominating that business. And with the performance advantages RISC chips offer, eventually they'll win out over X-86-based PCs."

I think about this and then ask, "How big is the performance advantage?"

"It depends on what new performance enhancement tricks we can come up with for the X-86. The RISC clock rate will always be higher, so we have to find other ways to improve our performance. We think it'll end up somewhere between 30 percent and 100 percent."

"Hmmm, that much. Well, obviously the smaller the gap, the better. But I believe that the advantages a customer receives by using an industry-standard PC are significant enough to offset a lot of performance. When we introduced the first 386 PC, it was accepted so quickly because it could use all the existing software and peripherals. Sparc workstations won't even come close to having that much flexibility. Some users who have extreme performance requirements might switch, but most users will continue to buy industry-standard PCs."

I pause to let that sink in, and then continue. "Think about it this way. When the PC market first started, there were several microprocessors in the race. Then IBM comes along and gives us all a gift by introducing a PC that can be cloned. And as a result of all our efforts, an industry standard develops and becomes dominant. The X-86 is a required component of the standard, which means Intel has won that race. As long as you continue to deliver competitive performance advances, you can't be removed from the throne. But if you make your top priority getting to a RISC chip that is competitive with the performance of the Sparc chip, it's like starting a new race where you may or may not be the winner. You should be doing absolutely everything you can to protect the industry standard and keep the X-86 advancing as rapidly as possible. You should protect your effective monopoly on the industry standard instead of making a new race your top priority."

I pause again and watch Grove think about this. After a while, he looks at Gates and says, "How much of a performance disadvantage can we have before a significant number of end users begin to switch over to Sparc?"

Gates replies, "A lot of the actual performance depends on the software, and it'll vary by the type of application. For most applications, a 30 to 50 percent performance disadvantage won't be a problem. I agree with Rod. As long as they don't have a sustained advantage of well over 50 percent, I don't think we'll lose many customers. And the ones we lose will be focused on a narrow range of applications in the workstation arena."

Grove adds, "The performance differential isn't a constant thing. When we come out with a new chip, most of the time we'll pass up their existing chip. Then when they come out with their next chip, it'll probably pass ours. So it's not like we'll be behind them all the time."

The four of us continue deeper into the details of performance and advantages of the industry standard. I try my hardest to explain why the industry standard has become so powerful. From my perspective, the pendulum swings in favor of my position for a while and then swings back in the other direction. As the meeting winds down, it is still unclear to me whether or not Grove will change his priorities.

Finally, Grove says, "I just can't get comfortable accepting the position of being at a performance disadvantage to a major competitor. Logically, I agree with your arguments. But I need to think about this and discuss it with my engineers. We can't afford to get this wrong."

I reply, "You're exactly right. None of us can afford to get this wrong. Please call me at any time of the day or night if you want to discuss it more."

As we leave the room, we all shake hands and agree to follow up soon.

When I arrived back in Houston, I still did not feel sure about what Intel was going to do. It was very uncomfortable to not be sure of something that could affect our company so much. I met with our strategy team shortly after returning to the office and told them how the trip had gone. The mood was somber.

Five weeks later, Barnes bursts into my office with a big grin on his face. He had just heard that Intel was reorganizing its processor development teams. Not only were they putting their top engineers on the 486 project, but they were also starting development on the 586 chip, which would eventually become known as the Pentium. Intel was continuing to work on a RISC processor, but not at the expense of either of the industry-standard chips. This was the best news that I could have imagined. I was thrilled as I contemplated what this would mean for Compaq's long-term success.

WE CONTINUED OUR TORRID PACE of new product announcements in 1989 with the introduction of the Deskpro 386/33 in May, two months ahead of IBM's comparable machine. As usual, we began shipping the product to our dealers ahead of the announcement. While several of our competitors, including IBM, had begun announcing new products several months before their shipments actually began in order to be viewed as "first" and as the "technology leader," none of them succeeded in getting large numbers of their new products to market ahead of us. Plus, the press had become wise to their practice.

The 33-megahertz 80386 chip was the last version of the 386 introduced before the 486 chip took over the performance leadership position late in 1989. We had succeeded in maintaining our performance leadership in 386-based computers throughout the entire 386 era. An article in *Computer Reseller News* on May 29, 1989, written by Alice Greene titled "Compaq Unveils 'A Screamer'" declared, "33 MHz System Cements (Compaq's) Lead In 386 Market."

On July 14, we signed a patent cross-licensing agreement with IBM. The company had first contacted Bill Fargo, our general counsel, to begin discussions shortly after the PS/2 announcement, but not much progress was made until September 1988. Then, just after the EISA announcement, IBM seemed to get upset and notified us that we were infringing many of its patents. At first I thought it was planning to use the tactic TI had used on us in 1982 and try to slow us down with a lawsuit. Instead, it was more like part of a larger program to extract some value for its intellectual property that competitors were using. It turned out that IBM was also infringing some of our patents. It was particularly concerned about our Roberts patent, which covered the dual-switching monitor Ken Roberts invented in 1982. David Cabello, our in-house intellectual property attorney, negotiated with the company for almost nine months before we finally reached an agreement. We paid IBM a one-time fee of $130 million that gave us unlimited access to all their patents filed through 1993. And it was much smaller than it would have been if we hadn't put a lot of effort into filing for patents on our own inventions.

Then, on October 15, 1989, we raised the bar for laptop computers by introducing the first "notebook" computer, so named because of its 8½ x 11-inch footprint, the same dimensions as a piece of notebook paper. In spite of their small size, the Compaq LTE and LTE/286 were full-function personal computers that employed several unique advances in technology. One of these, the first use of a 2½-inch-wide hard drive, enabled us to deliver up to 40 megabytes of storage.

We had made a strategic investment in Conner Peripherals, the company that invented this size drive and, as a result, were able to secure a six-month exclusive period on the tiny new component. We also received a six-month exclusive period for each new, higher-capacity drive introduced during the next three years. Thus, having defined a new category of laptop computer, we effectively owned that market segment for several years. As a footnote to history, Apple employed a similar strategy for its early iPods in 2001 and gained comparable results.

On October 17, 1989, an article in the *New York Times* written by Peter Lewis titled "Compaq Does It Again" opened with "Every once in a while a new computer comes along that leaps ahead of other machines in its class. This time, once again, it comes from the Compaq Computer Corporation."

True to our fundamental portable product philosophy, we had waited until the technology enabled us to deliver a "full-function" notebook computer. Contrary to analysts' predictions that we had missed the laptop market, the LTE and LTE/286 added to the strong start provided by our earlier SLT/286 and solidified our leadership position in the overall portable market.

One of the most visible signs of the rapid advance of technology in the '80s was the shrinking size and increasing power of portable computers. Every part of the computer changed dramatically over that seven-year period, and due in large part to the industry standard,

Evolution of Compaq Portables 1983–1989. The original Portable, Portable II, Portable III/386, SLT/286, and LTE/286.

many different companies contributed to those advances. The opportunity created by the rapid growth of the industry-standard segment of the PC market attracted most technology companies and provided the incentive for them to push themselves to beat their competition. And since they didn't have to adapt a new component to five or ten different systems, they could focus on the next advance while reaping the rewards of the previous one in the giant and rapidly growing industry-standard market.

The most important announcements of 1989 were yet to come. At the time of the original EISA announcement in September 1988, we had committed to launching our first EISA products by late 1989.

We would make good on that commitment, but there was a slowdown in the PC industry during the second half of 1989 that forced us to announce lower-than-expected earnings in the fourth quarter. The statement was released just a week before the launch of our new EISA products. We had become the darling of Wall Street with

We had become the darling of Wall Street with our financial performance over the last three years, and Wall Street did not kindly treat companies that let it down.

our financial performance over the last three years, and Wall Street did not kindly treat companies that let it down.

During the entire week before the announcement, the press was full of reports about our stock price getting battered. That really put a damper on the environment surrounding our upcoming product launch. This wasn't at all what I had in mind, but I pressed on, believing the announcement would be strong enough to put the earnings disappointment behind us.

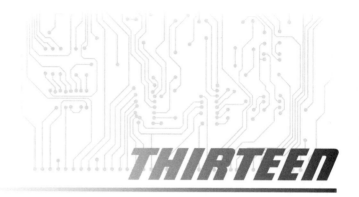

Exceeding Expectations

More than 1,000 people are gathered in the Houston Astrohall for what they think is Compaq's launch of our first 486 PC incorporating the EISA bus. The program starts out like one of our typical major product announcements, with Chairman Ben Rosen giving opening remarks.

"Today Compaq, the feisty company that has continued to recreate itself for seven years, is recreating itself once again."

As he completes his introduction, the hall is darkened and a dynamic laser light show complete with smoke and blaring music begins. The laser show subsides, then up comes a front screen filled with bold lettering: "UNLEASHING THE POWER." And on center stage is not a new PC like the audience is expecting, but an entirely new type of computer, the SystemPro.

I step to the podium. "The Compaq SystemPro combines a flexible system processor design enabling the user to take advantage of both 386 and 486 processor technology, including the capability to utilize multiple processors for unmatched performance; the 32-bit Extended Industry Standard Architecture expansion bus; Compaq Flex Architecture with Multiprocessing Support; and innovative, new, fixed-disk-drive array storage technology.

"As the industry leader in high-performance 386-based personal computers, Compaq is in the best position to migrate that technology to meet the expanding demands of departmental computing, networking, and multi-user applications. The System-Pro delivers superior performance at better prices than most minicomputers available today. We believe that its exceptional power and expandability make it a strong beginning for an entirely new class of personal computers, the PC system."

I list the details of each of its features, then describe the new software from Microsoft, SCO, Novell, and Banyan that is necessary to manage this advanced hardware in networking and multi-user environments.

We are aware that for many in the audience, details about "bits" and "bytes" are confusing. So I shift gears and begin comparing the SystemPro's price and performance with specific minicomputers sold by our major competitors. For the comparison, each computer was configured to serve as a host to sixty terminals in a multi-user environment using the UNIX operating system. I tell the audience that in comparison with the HP 9000 Series 835 minicomputer, the SystemPro is three times faster and costs $68,000 less, and it is six times faster and $135,000 less than DEC's VAX6310.

After finishing with the SystemPro, I introduce the DeskPro 486/25, Compaq's first PC based on the new Intel 486 chip. I explain in detail its features that are oriented toward advanced applications such as computer-aided design (CAD) and engineering workstations (EWS). The 486 delivers up to three times the performance of a 386 running at the same clock speed, which clearly moves it into competition with dedicated workstations.

I end by telling the audience that, unlike our normal practice, the 486 systems won't be available until early in 1990. During our testing of early production models, Compaq engineers discovered a bug in the 486 chip that Intel is now in the process of correcting. I didn't mention that one of our major competitors, in attempting to beat us to market, had already begun shipping 486 PCs with the bug in them.

After the presentations, attendees move to another room, where Compaq employees are demonstrating various networking and workstation configurations on more than a hundred computers. A significant number of staff from software and peripheral

The Compaq SystemPro, our first product using EISA,
outperformed minicomputers at much lower prices.

companies are also demonstrating their EISA products. There are even some IBM PS/2 Model 80s set up for comparison to the SystemPro. Demonstrations continue for several hours while attendees absorb the significance of these advancements.

We had gone beyond the obvious. As we had done so often, Compaq's strategy team had asked the right question: What customer needs can a computer with the EISA bus address? We realized that EISA could take the performance and capabilities of a 386- or 486-based computer into applications that PCs previously had not been able to address.

We began to look at functions traditionally performed by minicomputers and concluded that with new, higher-performance network interface controllers and disk-drive array controllers taking advantage of the increased performance capabilities of the 32-bit EISA bus, a 386-based computer could replace minicomputers in many workgroup applications. With a 486 processor, performance actually went beyond most minicomputers. And just to be absolutely sure we delivered the highest-performance EISA product, we decided to include an optional second processor that could be either a 386 or a 486.

We were aware of the problems IBM had created when it introduced the Micro Channel without any add-in boards available to demonstrate its performance potential. That's why we were committed to delivering a complete system, from Day One, that clearly showed the amazing performance EISA made possible. Compaq engineers worked with network hardware companies such as Novell to develop intelligent network controllers using the bus-mastering capabilities of EISA.

Other Compaq engineers who specialized in storage technology developed a disk-drive array controller that delivered four times the performance of non-arrayed drives, and therefore took advantage of the speed of the 32-bit EISA bus. After we combined dual processors with the intelligent network controller and the disk-drive array, we had a system that could win benchmark comparisons against minicomputers from HP, DEC, IBM, and others.

When I presented the results at the SystemPro announcement, it was almost too much for the audience to believe. Apparently, no one had ever seen a comparison between microprocessor-based systems and market-leading minicomputers. At first, it must have seemed like smoke and mirrors. We had anticipated skeptical reactions to our numbers and had hired the widely respected performance testing firm of Neal Nelson & Associates to conduct comparisons.

News reports of Compaq's new product announcements were generally positive, but somewhat muted due to our earnings warning the previous week. We were still projecting about 20 percent year-over-year quarterly revenue growth and net margin consistent with our model of 8 to 10 percent. The issue was simply missing expectations, but it was enough to take some of the gloss off our reputation for consistently beating expectations over the last three years. There was still some confusion over which bus would win in the long term, and now Compaq was adding another confusing claim that these new products could outperform minicomputers while costing much less. Plus, some people just couldn't accept that IBM had been beaten so badly.

A few reports did get the story right, however. Peter Lewis in the *New York Times* on November 7, 1989, wrote a story titled "Compaq Redefines High End." He noted how we "introduced two new products that redefine the high end of personal computer performance and are likely to change the landscape of office computing in the 1990s." Then, after reporting on the details of our new products, he concluded, "The debate over buses is a popular pastime in the engineering shops of IBM and its rivals, but we have yet to hear anyone say, in real life, 'Gee, that Micro Channel sure has made my life easier.' We suspect that most users do not care whether the data bus is ISA, MCA, the new EISA, or COAT (Chipmunks on a Treadmill), as long as the job gets done. The real effect of today's Compaq announcements will be felt in years to come, as desktop systems connected in networks replace the hulking minicomputers and mainframes that have been the backbone of office systems for 20 years."

Lewis was probably right; most users didn't care about the bus, but a lot of big companies did. If they stayed with the industry standard, they needed to be sure they would have access to the most advanced technology as it came to market. Conversely, if they were going to have to switch to Micro Channel–based PCs eventually, they wanted to minimize their investment in computers that would soon become obsolete. With the SystemPro announcement, we made it very clear that if they stayed with the industry standard, they would have access to advanced technology when they needed it.

In retrospect, none of the press or analysts saw the real significance of the SystemPro announcement. We had gone well beyond their expectations before, so it was easy to view the announcement as another event where Compaq had introduced a better product than their competition.

But this time we went far past merely exceeding expectations—this time, Compaq had called IBM's bluff and won. We proved that IBM had misled its customers and the industry when it said it would have to give up compatibility to get advanced performance. IBM did it again after the EISA announcement when it said it had looked at what the "Gang of Nine" had done and rejected it because it couldn't deliver the performance of the Micro Channel.

IBM had spun and stretched the truth all along for the obvious purpose of reducing or eliminating competition, but it could no longer get away with it. The SystemPro delivered technology advances way beyond anything that had even been discussed for the PS/2.

The SystemPro announcement effectively eliminated the only remaining logical argument for shifting to the Micro Channel. In the two and a half years since the original PS/2 announcement, essentially all of the customers who were going to shift to the PS/2 because of IBM's brand had already done so. It was clear to everyone that the Micro Channel part of the market was going to have higher prices because IBM had essentially eliminated any real competition. Customers also realized that once they had gone through the pain and expense of making the shift to Micro Channel, shifting back would be difficult. So if they were going to have to pay more for their PCs going forward, there needed to be a good reason for making the shift.

That reason had been, according to IBM, there would soon be desirable capabilities available on the Micro Channel that would not be available on industry-standard PCs. Since no such capabilities had emerged during the last two and a half years, many people were left wondering. But when Compaq demonstrated real, useful capabilities with EISA on the SystemPro that eclipsed anything available on the Micro Channel, the last significant reason for shifting to Micro Channel was eliminated.

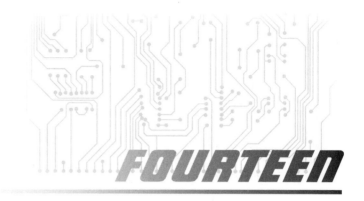

The Open Industry
Standard Wins

AFTER THE SYSTEMPRO ANNOUNCEMENT in late 1989, the outcome of the Micro Channel versus EISA war finally became clear. EISA, and with it the open industry standard, was going to win. It's difficult to put an exact date on when it won. Most wars end with the signing of a treaty, but there was no proclamation about the winner of this one. However, there would be no more serious challenges to the reign of the open industry standard throughout the remainder of the PC era.

Few people understood the real significance of the final battle. Many were aware of the "bus war" between IBM and the rest of the PC industry. Many more were aware of the war between IBM and Compaq for technology leadership of the PC industry and, later, for outright market leadership. But these were surrogates for a much more fundamental and far-reaching conflict, the outcome of which would determine the competitive nature and structure of the PC industry for decades to come.

Some would argue that the open industry standard was destined to win from the beginning. That it was so powerful and made so much sense that it could not be defeated. That rationale, however, grossly underestimates the power of IBM as a company and as a brand.

Before 1981, the PC industry had consisted of a diverse group of rebels and upstarts. Its leaders created relatively strong brands and viewed themselves as completely disconnected from the rest of the computer industry. But as soon as IBM entered the market in August 1981, they learned what a really strong brand could do. IBM quickly became the market leader, even though it had introduced a product made from off-the-shelf parts and showed nothing special except a label with "IBM" on it. Yet it remained in the PC industry leadership role for the next thirteen years.

> *Some would argue that the open industry standard was destined to win from the beginning...*
> *That rationale, however, grossly underestimates the power of IBM as a company and as a brand.*

Very likely, IBM would still be the PC industry leader today if it had fully understood the phenomenon that developed as a result of the popularity of its first product, and if it had chosen to lead the open industry standard instead of trying to end it. That would have been its strongest strategic move by far.

Most likely, though, IBM executives didn't believe there really was an industry standard. As far as they were concerned, a bunch of clones had copied IBM products. To be fair, the entire computer industry viewed it that way at first, calling it the "IBM-compatible market" and even the "IBM standard." By late 1983, however, I had introduced the term "industry standard" and later, "industry-standard architecture." If IBM executives noticed at all, they probably viewed these terms as marketing hype or our wishful thinking.

As is often the case, IBM's great success led to an overconfidence that caused it to miss what was really happening. IBM viewed the rise of the clone PC companies as the same result that had occurred in the past, when it failed to protect one of its products with enough proprietary technology. So IBM naturally concluded the solution was the same one it had successfully used every other time: come out with a replacement product that is more protected and quickly put an end to the clones.

Initially it positioned the PS/2 and Micro Channel so there could be absolutely no cloning. But fairly quickly, IBM backed off and said it would license Micro Channel to the clones for a license fee and a royalty of 5 percent of sales. Since many of the clones made less than that in profits, the effective result was the same as not allowing cloning. However, this way looked better, especially to the United States Justice Department.

IBM could have continued leading the industry long term if it had chosen to introduce something like EISA instead of Micro Channel. Compaq and the others would have had no choice but to follow, even if we had to license the 32-bit part of the bus. There would have been no "what IBM should have done" alternative for us to use against it. And adding a strategic move that made so much sense to its already incredible brand would have made it seem even more invincible.

But that was not the way IBM executives thought. They must have believed they could get all the clones to follow them to the Micro Channel; after all, the clones had done so with every major technology advance up until then, except one. IBM unintentionally had given Compaq the opportunity to introduce the first 386 PC ahead of it. When Compaq's 386 PC became a great success, it clearly demonstrated there were many customers willing to follow a company other than IBM to a major new processor, as long as it was compatible with the industry standard. But instead of worrying about this inconvenient fact, they probably assumed that the PS/2 would crush Compaq's 386 and prove that getting out in front of IBM was a mistake.

Even though introducing the proprietary—but incompatible—Micro Channel wasn't their best strategic move, IBM still could have succeeded if it hadn't let Compaq get ahead with the 386. Ironically, we wouldn't have had the reputation and credibility to successfully organize and lead the EISA coalition if we hadn't succeeded with the first 386 PC and demonstrated the importance of backward compatibility with the industry standard. In addition, if IBM had made it less costly for clones to license the Micro Channel by charging a more palatable royalty of, say, 1 percent, IBM would likely have lured the clones to follow it, thereby retaining tight control of the technology and, therefore, the industry. If it had put both of these together, IBM would almost certainly have succeeded in eliminating the open industry standard and continued its reign for a very long time.

When IBM first introduced the PS/2, almost everyone focused on the impact the computer would have on the clones. It was correctly believed that IBM was driven by a desire to reduce competition and regain the significant market share it had lost to the clones. What was generally overlooked was the potential impact on Intel and Microsoft if the PS/2 had become the new standard.

While IBM was jointly developing OS/2 with Microsoft, there is no doubt that its long-term goal was to completely control the operating system, effectively cutting Microsoft out of the picture. Without the cash flow from its position as sole supplier of the industry-standard operating system, Microsoft wouldn't have been able to persist in pursuing—and eventually taking the lead in—several other key software market segments.

Intel's position as the dominant supplier of processors for industry-standard PCs was threatened by virtue of it having granted IBM manufacturing rights to the 286 chip as part of an equity investment in the early '80s. It was reported that IBM was planning to use its own

286 chips in the PS/2, but due to chip manufacturing problems, it had temporarily backed off. If it had succeeded in making the PS/2 the standard, it would have been in a position to gradually cut Intel out of the picture as well.

One plausible explanation for why IBM lagged a year behind Compaq in delivering a 386 PC was that it may not have planned on using the 386 at all. If the PS/2 had become the standard using its 286 chip, IBM could have easily introduced a proprietary next-generation processor instead of the 386. This would have completely cut Intel out of the picture and put IBM in the position of supplying processor chips to all Micro Channel licensees.

Once EISA was established as the industry-standard bus, the "Gang of Nine" took control of future changes in system architecture. With IBM no longer in a dominant position, Intel alone controlled advances in the microprocessor; it no longer had to worry whether IBM would change to a different processor in its next product. And so long as each new Intel processor was compatible with the previous one, it would automatically be accepted as the new industry-standard microprocessor. The same was true for Microsoft and its operating systems. As a direct result of our success in defeating IBM's Micro Channel, each company gained unfettered monopoly positions in the industry standard for the long term.

Compaq certainly gained a lot as well. But when we succeeded in keeping the playing field level by saving the open industry standard, we ensured that we would have to play on that level playing field too. We knew how to do that better than anyone, and as a result continued to thrive. Compaq's management team had grown up in the industry-standard environment. We had matured and gelled as a team, and in my opinion were one of the best in the industry.

Compaq Management team, early 1990. Clockwise from left: Daryl White, Jim Eckhart, Bill Fargo, Murray Francois, Bob View, John Gribi, Kevin Ellington, Gary Stimac, Jim Harris, Mike Swavely, Ben Rosen, Rod Canion, and Eckhard Pfeiffer.

AFTER TEN YEARS AT THE HELM, I left Compaq at the end of 1991. Parting was bittersweet. In truth I was somewhat burned out by the intense, nonstop pace of those ten years, especially during 1991. That year we were caught off guard by an economic recession, by six of our top ten dealers merging into three, and by an unexpected strengthening of the dollar. The result was Compaq's first quarterly loss since 1983 and our first layoffs ever. Laying off hundreds of dedicated people weighed very heavily on me.

Jim left Compaq at the same time. Jim had led the design of the original Portable and built an engineering team that was second to none. He established the design discipline in Compaq that was fundamental to achieving and maintaining our reputation for quality and ruggedness. Jim was also my sounding board and advisor throughout the entire ten years.

But before we left we had to deal with the question of strategy. I took the strategy team off-site several times during the summer and early fall of 1991 to carefully analyze our strengths and weaknesses, and decide if and how we needed to change. We decided we could no longer stay with just high-end products and prices. It became clear that we could leverage one of our most valuable assets, our strong upscale brand, to help enter the low end of the market. So in late summer we started a "crash" project to design a low-cost computer that still had differentiating features.

Eckhard Pfeiffer took over as CEO when I left and led Compaq to launch the low-price product line in June 1992. He and his team did an amazing job of taking the basic, low-cost design from the "crash" program and parlaying it into an incredible total of sixteen new products. The new machines included: low-priced desktop PCs, Compaq ProLinea; low-priced notebook PCs, Contura; and upgradable desktop PCs with advanced graphics and audio capabilities.

The announcement sent shock waves through the industry worldwide. No one expected Compaq to make such an aggressive move and to have such aggressive pricing. The strength of the low-end product launch shifted momentum back to Compaq, and its sales growth accelerated again. Revenues nearly tripled over the next three years, growing from $4.1 billion in 1992 to $10.8 billion in 1994. In that pivotal year, Compaq became the leading supplier of PCs worldwide, while IBM quietly announced that it would no longer be producing the PS/2.

As the tides ebbed and flowed, in 2001 Compaq achieved sales of over $33 billion and merged with its old ally from the PC wars, Hewlett-Packard. Another economic recession had come along and both companies were struggling. The little company that wouldn't be pushed around by IBM had run its course, as the combined company took the Hewlett-Packard name. The Compaq brand today lives on in a somewhat subdued role as HP's consumer PC product line.

Compaq's legacy also lives on in its impact on the PC industry and all our daily lives. The many advantages of the industry standard are still with us, giving us broad choices of PC features and brands at incredibly low prices. Just as importantly, if Compaq hadn't succeeded in stopping IBM and preventing Intel's loss of its monopoly position in processors, Intel wouldn't have earned such enormous profits through the '90s and, as a result, wouldn't have been able to advance chip and processor technology as rapidly. If that hadn't happened, we almost certainly would not have the technology we have today that drives the smartphones and tablets we all enjoy.

Epilogue
How Apple Became the Computer Industry Leader

APPLE WAS THE ONLY major personal computer company to resist joining the PC open industry standard and survive. It carved out a niche in the late '70s and early '80s and attracted a very loyal group of customers. Eventually, though, even Apple had to enable its Macs to run industry-standard software. Even so, it was reportedly on the verge of bankruptcy by the late '90s.

Then Apple discovered a paradigm shift ahead of its competitors and surged into the leadership position. It didn't do it by introducing a better Mac or PC; it simply wasn't possible to overcome the strength of the open industry standard. Apple did it by inventing an entirely new way of using the Internet.

The iPad wasn't the result of a brilliant idea conceived out of thin air. Rather, it was the result of following a logical progression starting from the original "seed"—the iPod—along with some very innovative new technology. Like many paradigm shifts, the change was invisible to the establishment until it was so far behind it was too late to catch up.

When Apple first introduced the iPad, tablet computers with touch screens had been around for over a decade. At some point, every successful PC company had introduced a tablet and thought it had found the right formula. They all tried, and they all failed. No doubt there were some chuckles in the executive ranks of those companies when Apple introduced yet another tablet computer. There simply wasn't a market for such a product, these execs thought—but they were dead wrong.

In one sense, they were right: There wasn't a significant market for another tablet computer similar to those they had all tried. But the iPad was a totally different animal, essentially a wolf in sheep's clothing. The first signs of this difference appeared in 2008, two years before the iPad, when Apple introduced the second-generation iPhone. To be fair, no one, including Apple, connected the smartphone market to the PC market at that time. And when the iPad was introduced in 2010, almost no one saw it as a threat to PCs because it couldn't replace all the functions of a laptop.

Back at the beginning, the iPod had no real connection to the computer business, but it was an interesting market opportunity that offered some additional revenue and profit to a struggling computer company. Several other companies, including Compaq, had introduced competing products, but Apple got it right by taking a page out of the playbook Compaq used for the first notebook computer in the late '80s. Apple decided that small was the key, and bought up all the smallest hard drives available at the time. No other company was able to deliver a music player as small as the iPod because none of them could get enough small disk drives.

The really important innovation was not the iPod itself, but the iTunes online store that followed. Coming from a very proprietary mind-set, Steve Jobs wanted to sell all the music played on iPods. There was a lot of trial and error in the beginning, but eventually Apple got it right. The combination of the smallest music player and the convenience of the iTunes online store was strong enough to overcome significant customer resistance to being locked into a single source. Although Apple became the clear market leader, the iPod and iTunes were still just a nice source of revenue.

Then in 2007 Apple introduced its first smartphone. The iPhone had significant limitations, including the inability to wirelessly synchronize with Microsoft Exchange and no third-party apps, but it did include an innovative new capability: the multi-touch user interface. While most people viewed this as merely a "cool" feature, a few realized it was the solution to a previously unsolvable problem. The minimum usable screen size in notebook computers was generally accepted to be between 11 and 13 inches on the diagonal. Any smaller and text was either too small to read or the screen couldn't show enough information. A smaller screen became a lot more useful when users could expand and shrink any part of the screen instantly with the touch of two fingers.

In hindsight, Apple was probably lucky the first iPhone didn't get it all right. The company didn't have the "can do no wrong" aura it has today; if it had, competitors likely would have paid more attention and realized what they were dealing with. As it happened, many weren't expecting Apple to be a significant player in the smartphone market. The shortcomings of the first iPhone simply confirmed its competitors' beliefs. A year later, when the second iPhone was introduced, most of Apple's competitors still weren't worried about it and paid no attention. But this time Apple not only had a good smartphone, it had put the final piece of the puzzle in place to create an entirely new market.

That final piece was third-party programs known as "apps." Apps had been around for a long time on smartphones using the Palm or Microsoft platforms, so the fact that the second iPhone could use third-party apps didn't attract much attention. It was the addition of apps to the iPhone's other features that created Apple's breakthrough.

Six key elements had to converge for the breakthrough to occur:

1. Multi-touch user interface
2. Third-party apps
3. Continuous connection to the Internet
4. Instant "on"
5. Competitive phone capability
6. App distribution through the iTunes online store

The fifth and sixth elements weren't really part of the iPhone's breakthrough functionality, but they were necessary to get the broad acceptance required to attract millions of users and thousands of app developers. In the beginning, many customers were just looking for a good smartphone and picked Apple's because it was a good phone and had its cool multi-touch feature.

If the iTunes store hadn't already been in place, it would have taken much longer for the large number and wide variety of apps to develop, thereby reducing or eliminating Apple's lead over its competitors. Jobs couldn't have known when Apple created the iTunes store in 2001 that it would be so important to enabling individual programmers to get their innovative iPhone apps into broad distribution so easily.

The first four elements added together created a new way to access the Internet's products, services, and information. As soon as hundreds of apps became available through the iTunes store, users began to glimpse the future of Internet access. In literally seconds, they could check the current weather forecast or find movie listings and show times at a nearby theater. When the thought occurred to a user, they would simply turn on the iPhone instantly, touch the screen to select a particular app, and then touch the screen once or twice more to access the information they desired. The technology was simple enough for almost any human, and so fast that it became the preferred way to access the Internet.

When Apple made it easy for app developers to create and sell their creations, it unleashed an army of tens of thousands of individual innovators to provide solutions to real user needs. It was an interesting parallel to spreadsheet programs running on PCs in the '80s. Spreadsheets didn't just do numerical calculations; they included a language so simple users didn't even realize they were writing programs. Once this tool was in the hands of tens of thousands of unwitting "programmers," solutions were found to thousands of real user needs. As a result, the PC became much more useful and market growth accelerated.

With the easy distribution of apps through the iTunes store, Jobs duplicated one of the important advantages enjoyed by industry-standard PCs in the '80s, effectively unleashing an army of solution creators. And it wasn't just the iPhone that could use all these apps. A relatively intuitive parallel step for Apple was to create the iPod Touch, essentially the iPhone without the phone. The iPod Touch performed all the same functions as the iPhone, including running the same apps, but by using a Wi-Fi connection to the Internet instead of a wireless phone data network. The places a person could use it were reduced, but an iPod Touch user didn't have to pay a phone bill or commit to a two-year contract. Within a year, millions of adults were using the iPhone and millions of kids were using the iPod Touch; a few years later, more than 60 million people were using "i-devices" to interact with the Internet.

Still, many computer executives viewed the iPhone as a phone and gadget, unrelated to PCs. They didn't realize they were already a long way behind Apple in the next important computer market.

The main thing Apple needed to do to ensure instant success for the iPad was to make it work like the iPhone and run the 150,000 iPhone apps that were available in the iTunes store. Apple did that. When the iPad was introduced in March 2010, sales took off like a rocket. Sixty million people already knew how to operate an iPad because they owned iPhones and iPod Touches. Apple had created a giant pent-up iPad demand that they were unable to fill for months. In spite of the incredible number of units sold, many computer company executives still thought of the iPad as just a big iPhone. It wasn't clear to them how a big iPhone was going to be a problem for PCs.

In early March 2010, just before the iPad was announced, I began contacting executives of several PC companies to discuss what the iPad was going to mean. I'm sure they all wondered why I was doing this, and what I wanted. They had difficulty accepting that I wanted to share my insights with them and help them avoid missing an important opportunity. I tried to convince them the iPhone and iPad were together going to become the preferred way of accessing Internet goods, services, and information. While iPads wouldn't replace notebook PCs altogether, I explained that they would attract many potential new PC buyers, because the main thing many consumers wanted was Internet access.

I told them that to effectively compete with Apple, they needed to strategically address the customer needs Apple was filling, and that it wasn't just a hardware issue. In each of those conversations, I could sense that the executives thought they had the situation under control. During one particular meeting, out of frustration, I told them they were so far behind Apple they could barely see their taillights. I was promptly shown the door.

It wasn't long before PC companies began to see their sales growth diminish and then, a while later, their sales decline. The executives were correct in believing that neither smartphones nor tablets could replace all the functions of a notebook PC, but they were very late, unfortunately, coming to the realization that the next major source of growth for PC companies was Internet access. Smartphones and tablets together simply did a better job of addressing that market than notebooks.

As successful as Apple has been in the smartphone and tablet markets, it could have been even more dominant if it hadn't made two key mistakes. Jobs allowed Eric Schmidt, then Google's CEO, to remain on Apple's board of directors during the early stages of iPhone development. Jobs must have articulated his vision for iPhones and iPads, which set in motion events that led to the establishment of Apple's most serious competitor.

Google subsequently developed the Android operating system for smartphones and tablets. In a move similar to Microsoft's in the '80s PC market, it then began providing its system to all Apple's competitors and potential competitors. Google was trying to create an industry standard much like the PC industry standard, no doubt hoping for the same powerful results.

Apple's second major mistake compounded the effect of the first. Apple introduced the first iPhone for exclusive use on the AT&T wireless network in the United States. That made sense because Apple, as an unproven phone provider, needed to get into broad distribution. But the mistake was remaining exclusively with AT&T for over three years. Apple apparently didn't realize the power that wireless carriers had over their customers.

During the second and third year of iPhone sales, Apple's apps were far superior to the Android's. If the iPhone had been available to customers of all other major wireless carriers, Android phones from Samsung, HTC, Motorola, and others wouldn't have taken off as they did. When a customer had to pay several hundred dollars to get out of a contract and switch to AT&T in order to have the iPhone, they usually ended up staying with their provider and looking for another choice similar to the iPhone. That gave Android-based phone providers an incredible opportunity to gain a foothold in the market. It wasn't long before the Android apps improved significantly. By the time Apple began to slowly add other wireless carriers, Android-based phones had become widely accepted and sold many millions, eventually surpassing Apple's iPhone sales.

Proof that Android's success didn't have to happen is seen in the tablet market. Android-based smartphone companies probably assumed they could achieve success with tablets the way they had with smartphones, but initially, Android-based tablets flopped. Why? There weren't any contracts with wireless carriers to inhibit customers from choosing the best tablet available, which for a long time was clearly the iPad. Since the iPad had superior apps, tablet buyers almost always bought the Apple product, even if they were already using an Android-based smartphone. It would have been the same result with smartphones if the iPhone had been available on every carrier early on.

At this time, Samsung looks like the company best positioned to give Apple its most competition. It got there by filling a lot of niches that Apple hadn't addressed. They seem to understand how to build a strong brand and take advantages of Apple's mistakes.

In late 2012, Microsoft finally entered the fray with the introduction of Windows 8, Windows 8 RT, Windows Phone 8, and the Surface tablet. With regard to the smartphone and tablet markets, Microsoft had long been considered to be in a coma, if not already dead, but it has usually been a mistake to count Microsoft out too early. It continues to have a dominant position in the corporate market for industry-standard PC operating systems and applications. Apple has made significant inroads into the corporate world with the iPhone and iPad, but it's still not too late for Microsoft to hold on to a significant share of that market if it can make smartphones, tablets, and PCs work together in a seamless, useful way.

Surprisingly, Apple gave Microsoft an opening by not migrating its multi-touch technology to its notebook screens, something it could have easily done. Microsoft took advantage of that opportunity by including touch-screen notebooks and desktops with multi-touch capability as part of its Windows 8 introduction. It is the first company to enable smartphones, tablets, and PCs to operate exactly the same way.

Microsoft made one major mistake in introducing its original Surface RT tablet. All Microsoft had to do to significantly increase its chances for success was follow Apple's lead and make sure the Surface RT ran the tens of thousands of apps available for Windows Phone 7 and 8. If it had, initial buyers of Microsoft's first tablet would have had immediate access to many important apps, just as initial buyers of the iPad did. But Microsoft didn't do it that way, and as a result, the Surface RT has been very slow to gain those important apps because of poor sales. As with the "chicken-or-egg" question, poor sales may continue because it doesn't have those apps.

Microsoft also made the first version of Windows 8 more difficult to use than it should have been. It will take a while before PC users get completely comfortable with multi-touch on Windows 8 PCs, and some tweaking on Microsoft's part will be required to get it right. But multi-touch capability is so intuitive and powerful that users are certain to prefer it on all their devices. Even if Apple responds by including multi-touch screens in future notebooks, Microsoft can still hold on to a leading share in the corporate market if it gets Windows 8 right.

Apple has taken over the leadership position in computers not by beating the open industry standard—but by going around it and inventing the next big thing. There is no doubt that the visionary leadership of Steve Jobs enabled its success. Many now wonder if Apple can continue to lead the market without Jobs. Apple has many talented and innovative people, including Tim Cook, its new CEO. Very likely, it will continue to introduce a steady stream of innovative new products into the market. But with or without Steve Jobs, the long string of important innovations that began with the first iPod, ended with the third-generation iPad. Each step along that path moved

Apple farther and farther out in front of its competition in the smart-phone and tablet market and enabled it to build an incredible reputation. But as it was with Compaq at the end of the '80s, that string of successes couldn't continue unbroken forever.

The iPad Mini was an important introduction because it addressed a niche that Apple had left for its competitors to fill. They need to do more of that. But for the foreseeable future, each new smartphone or tablet product will be an incremental step, not a big differentiator. Android and Windows 8-based products have a chance to close the competitive gap, but it depends on how well they plan and execute their strategies.

It is unlikely, however, that anyone will knock Apple out of their leadership position any time soon. But you never know—HP, Dell, Samsung, and others have the potential to make significant gains, if they understand customer needs and formulate strategies to address those needs.

Just as it was with PCs in the '80s and '90s, the real beneficiary of all this competition is the consumer: you.

Appendix
Compaq Timeline

1982

JANUARY 8	Idea for IBM PC-compatible portable computer
FEBRUARY 16	Incorporate as Gateway Technology, Inc.
FEBRUARY 22	Close first venture capital funding of $1.5 million
MARCH 19	Meet with Bill Gates to ask for IBM-compatible MS-DOS
JUNE 7	National Computer Conference (NCC) in Houston
	Prototype shown to dealers, investors, and the press
SEPTEMBER 9	Close second venture capital funding of $8.5 million
OCTOBER 12	Decision to distribute only through computer dealers

NOVEMBER 4	Change name to Compaq Computer Corporation
	Announce Compaq Portable PC at New York press conference
DECEMBER 31	Employees: 100 Dealers: 50 Sales: $0

1983

JANUARY 31	First month of shipments: 250 Portables
FEBRUARY 7	Decision to plan for annual sales revenue of $100 million
MARCH 24	Close third round of equity financing of $20 million
OCTOBER 25	Announce Compaq Plus with first shock-mounted hard drive
DECEMBER 9	Compaq Initial Public Offering raises $66 million
DECEMBER 13	Decision to enter desktop PC market and start "Bullet" project
DECEMBER 31	Employees 600 Dealers 959 Sales $111 million

1984

JANUARY 16	Decision to continue full production, despite few dealer orders
JANUARY 24	Apple announces first Macintosh
FEBRUARY 16	IBM announces Portable PC

APRIL 20	Compaq announces record first quarter despite IBM Portable
JUNE 28	Compaq enters desktop PC market with Deskpro
AUGUST 15	IBM launches PC-AT, first 286-based PC
SEPTEMBER 4	Decision to wait for 8-megahertz 286 to leapfrog AT performance
DECEMBER 31	Employees 1,318 Dealers 2,034 Sales $329 million

1985

APRIL 30	Compaq introduces Deskpro 286 and Portable 286, 33 percent faster than IBM's AT
SUMMER	IBM and clones cut prices
DECEMBER 6	Compaq moves company listing to NYSE
DECEMBER 31	Employees 1,838 Dealers 2,806 Sales $503 million

1986

FEBRUARY 20	Compaq introduces Portable II, smaller and lighter 286-based portable *Fortune* magazine announces Compaq reaches Fortune 500 in shortest time ever
MARCH 12	Decision to introduce 386 PC ahead of IBM
SUMMER	More IBM and clone price cuts
SEPTEMBER 9	Compaq announces Deskpro 386, first 386-based PC

| DECEMBER 31 | Employees | 2,209 |
| | Sales | $625 million |

1987

FEBRUARY 17	Compaq announces Portable III: 18 pounds with flat screen
APRIL 2	IBM announces PS/2 product line with Micro Channel 32-bit bus
APRIL 16	Compaq execs begin "New Coke" publicity battle to fight the PS/2
MAY 18	Decision to reverse engineer the Micro Channel
MID-AUGUST	IBM ships its first 386, PS/2 Model 80
SEPTEMBER 29	Compaq launches Deskpro 386/20 and Portable 386, 2X PS/2 Model 80 speed
OCTOBER 28	Decision to develop a compatible 32-bit bus for the PC industry
DECEMBER 31	Employees	4,052
	Sales	$1.2 billion

1988

JANUARY 4	Decision to halt Micro Channel reverse-engineering project
JUNE 20	Introduction of Deskpro 386/25 and Deskpro 386s, first 386sx PC
JUNE 22	HP joins Compaq, Microsoft, and Intel to form Extended Industry Standard Architecture (EISA) Coalition

SEPTEMBER 13   "Gang of Nine" press conference announces EISA coalition

SEPTEMBER 19   Compaq introduces the lower-priced Deskpro 386/20e

OCTOBER 17   Compaq enters laptop segment with SLT/286

DECEMBER 31   Employees   6,503
              Sales $2.1 billion

1989

FEBRUARY 21   Compaq terminates relationship with Businessland

MAY 22   Compaq introduces Deskpro 386/33, using the last version of 386 chip

JULY 14   Compaq signs patent cross-licensing agreement with IBM

OCTOBER 15   Compaq introduces first notebook PCs, the LTE, and LTE/286

NOVEMBER 6   Introduction of SystemPro and Deskpro 486/25, the first EISA products

DECEMBER 31   Employees   9,539
              Dealers   3,300
              Sales $2.9 billion

1990

MARCH 5   Compaq introduces Deskpro 385/25e

MAY 21   Compaq introduces Deskpro 286n and 386n

JUNE 18   Compaq introduces Deskpro 386s/20 and SLT 386s/20

JULY 23	Compaq introduces Deskpro 386/33L, 486/33L, and SystemPro 486
OCTOBER 15	Compaq Introduces LTE 386s/20
DECEMBER 31	Employees 11,420
	Sales $3.6 billion

1991

OCTOBER 23	Compaq announces $70 million loss and layoffs
OCTOBER 24	Canion and Harris leave Compaq, Pfeiffer becomes CEO
DECEMBER 31	Employees 10,059
	Sales $3.3 billion

1992

JUNE 15	Compaq announces 16 new low-price products
DECEMBER 31	Employees 9,559
	Sales $4.1 billion

1994

MID-YEAR	IBM stops producing the PS/2
DECEMBER 31	Employees 14,372
	Sales $10.8 billion

Compaq becomes the leading PC supplier worldwide

Market Share, units worldwide:

Compaq	10.3 percent
IBM	8.5 percent
Apple	8.5 percent

2001

SEPTEMBER 3	Compaq and HP announce definitive merger agreement
DECEMBER 31	Sales $33.6 billion

Market Share, units worldwide:

Dell	13.3 percent
Compaq	11.1 percent
HP	7.2 percent
IBM	6.4 percent

Acknowledgments

I WOULD LIKE TO THANK Russ Setzekorn for his considerable help as sounding board, researcher, writer, and editor in helping make this book a reality.

I also want to thank my colleagues from Compaq who helped and supported me in many ways: Hugh Barnes, David Cabello, Mike Clark, Wayne Collins, Ross Cooley, Kevin Ellington, Bill Fargo, Steve Flannigan, Kim Francois, Murray Francois, John Gribi, Jim Harris, Bill Murto, Greg Petsch, Ken Price, Bob Stearns, Gary Stimac, Mike Swavely, and Bob Vieau.

I am very appreciative of the invaluable assistance from my publishers, Glenn Yeffeth and Debbie Harmsen; my agent, Bill Gladstone; and my five editors in creating the final manuscript, not to mention teaching me the writing, editing, and publishing processes.

Also, special thanks to my wife, Cam Canion, who supported me throughout this journey and turned out to be one of my most helpful editors.

Index